THE PEACE IN PERIL

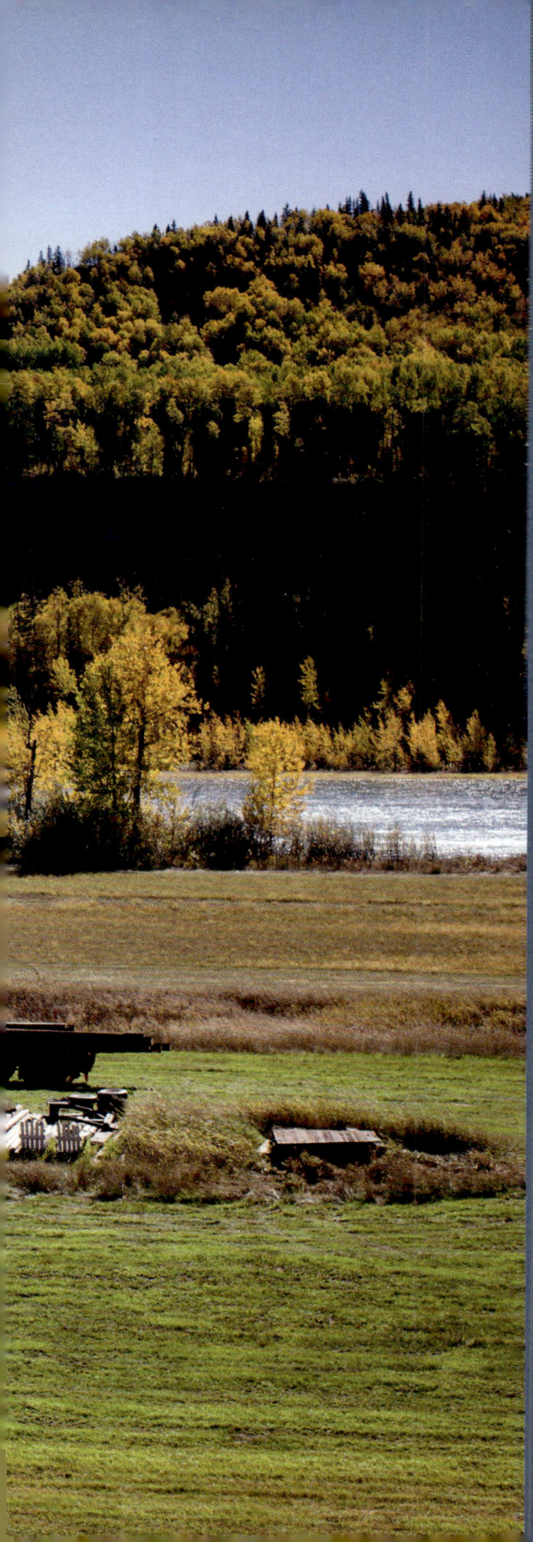

THE PEACE IN PERIL

The Real Cost of the Site C Dam

Christopher Pollon
with photographs by Ben Nelms

Previous pages: The site of the dismantling of Leo Rutledge's log-cabin-styled barn.

Opposite: Caroline and Derek Beam's horses are pictured on their land in Hudson's Hope on the Peace River.

1 2 3 4 5 — 20 19 18 17 16

Harbour Publishing Co. Ltd.
P.O. Box 219, Madeira Park, BC, V0N 2H0
www.harbourpublishing.com

Edited by Joanna Reid
Indexed by Kyla Shauer
Cover design by Anna Comfort O'Keeffe
Text design by Roger Handling
Printed and bound in Canada

Canada Council for the Arts Conseil des Arts du Canada

Harbour Publishing acknowledges the support of the Canada Council for the Arts, which last year invested $153 million to bring the arts to Canadians throughout the country. We also gratefully acknowledge financial support from the Government of Canada through the Canada Book Fund and from the Province of British Columbia through the BC Arts Council and the Book Publishing Tax Credit.

Library and Archives Canada Cataloguing in Publication
Pollon, Christopher, 1967-, author
The Peace in peril : the real cost of the Site C dam / Christopher Pollon ; Ben Nelms, photographer.

Includes bibliographical references and index.
Issued in print and electronic formats.
ISBN 978-1-55017-780-0 (paperback).--ISBN 978-1-55017-781-7 (HTML)

1. Dams--Environmental aspects--Peace River (B.C. : Regional district). 2. Water resources development--Peace River (B.C. : Regional district). I. Nelms, Ben, 1988-, photographer II. Title.

TD195.D35P65 2016 333.91'4140971187 C2016-906240-6 C2016-906241-4.

CONTENTS

ACKNOWLEDGMENTS

THIS BOOK WOULD NOT EXIST without the generosity, insights and contributions of: Arno Kopecky, David Beers, Ken and Arlene Boon, Harold Steves, Michael Harris, Erin Millar, Deborah and Ross Peck, Tom Sandborn, North Peace Search and Rescue, Marvin Shaffer, Craig Benjamin, Mike Wilson, Stan Persky, Art and Laurel Hadland, Wendy Holm, Eveline Wolterson, Vernon Ruskin, Emma Gilchrist and *DeSmog Canada*, Don Hoffmann, the staff at the Hudson's Hope Museum and Archives, John Werring, Harry Swain, Roland Willson, Marc Eliesen, Ben Parfitt, Tyee Bridge, Lynne Hill, Jordon Tomblin, Jane Calvert, Don Hoffmann, Jeff Galius, Matthew Nefstead, Jack Woodward, Jonathan von Ofenheim, Montana Cumming, Luke Gleeson, Jay Sherwood, Michael Church, Mark Jaccard, Mike Van Zandwyk, Vic Gouldie, David Hughes, Robin Banergee, Michelle Hoar, Caroline Beam, Roger Bryenton, Adam Reaburn, Scott Powell, Moja Coffee and Prado Cafe (where the idea for this book was born). Immense thanks also to the extraordinary people at Harbour Publishing, including Anna Comfort O'Keeffe and Brianna Cerkiewicz, and my editors Joanna Reid and Pam Robertson. A special thanks to my girls, Glenna and Gabrial Pollon, and the Ontario Pollons.

—Christopher Pollon

Previous pages: Two men fish near Lynx Creek on the Peace River east of Hudson's Hope.

THANK YOU to the Nelms and Flanagan families for supporting this project, and special thanks to Darryl Dyck, Cole Burston and Aaron Fife. I am also eternally grateful to my loving wife Mariel, for always keeping an eye on me.

—Ben Nelms

Opposite: Brittle sedimentary rocks are ubiquitous on the islands that dot the Peace River between Hudson's Hope and Taylor.

MAP OF PEACE RIVER CANOE TRIP EXPLORING FUTURE SITE C DAM FLOOD AREA

Legend

- Site C reservoir
- Dawson Creek/Chetwynd Area Transmission Project
- 1 Start of trip
- 2 Caroline Beam's home
- 3 The Gates
- 4 Night 1 camp
- 5 Leo Rutledge's former homestead
- 6 Ross and Deborah Peck's home
- 7 Night 2 camp
- 8 Ken and Arlene Boon's homestead, PVEA fundraiser location and Bear Flats
- 9 Night 3 camp
- 10 Charlie Lake cave
- 11 Rocky Mountain Fort archeological site
- 12 BC Hydro construction site
- 13 End of trip

Williston Lake

W.A.C. Bennett Dam

Dinosaur Reservoi

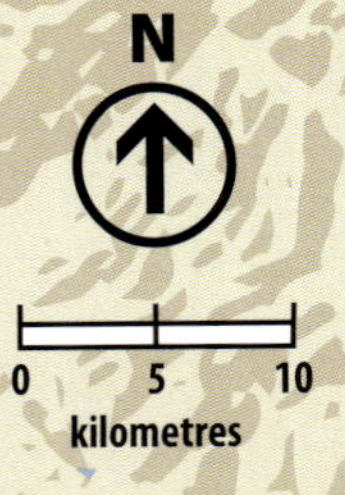

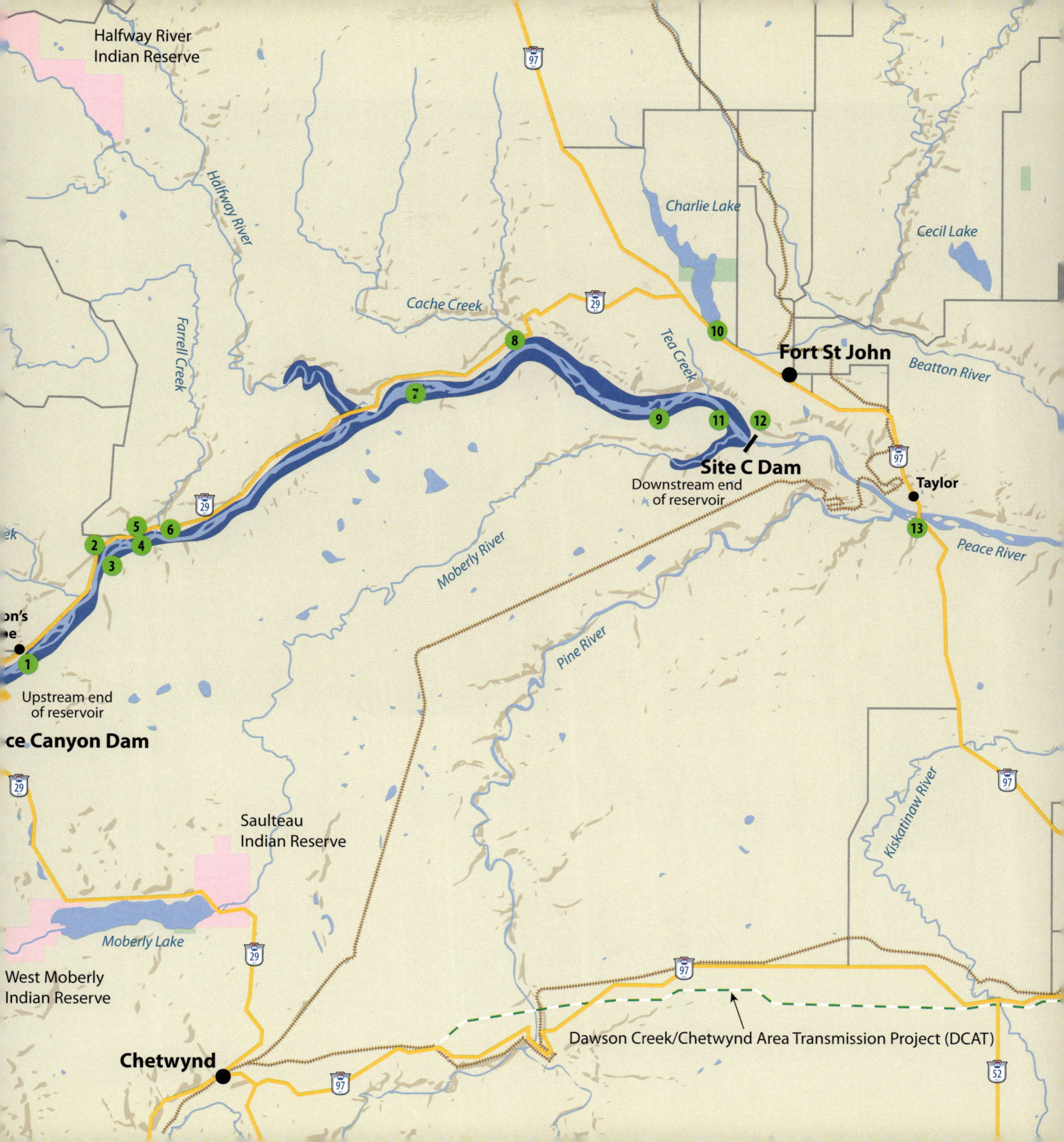

Halfway River Indian Reserve
Halfway River
Farrell Creek
Cache Creek
Charlie Lake
Cecil Lake
Tea Creek
Fort St John
Beatton River
Site C Dam
Downstream end of reservoir
Taylor
Peace River
Moberly River
Pine River
Upstream end of reservoir
ce Canyon Dam
Saulteau Indian Reserve
Moberly Lake
West Moberly Indian Reserve
Kiskatinaw River
Chetwynd
Dawson Creek/Chetwynd Area Transmission Project (DCAT)
97
29
52
1
2
3
4
5
6
7
8
9
10
11
12
13

INTRODUCTION: INTO A VALLEY OF GHOSTS

Opposite: A moose sits on the side of Highway 29, five kilometres east of Hudson's Hope.

MONDAY MORNING RUSH HOUR just north of Prince George, British Columbia. As we drive up Highway 97 toward Chetwynd and into the Peace River Country, a southbound pickup truck passes at an intersection. The windows are tinted, but the occupant in the back is still visible: a severed moose head with antlers so immense they fill the entire cab. Bloodshot eyes stare skyward as if to plead, *Why me?*

Moose meat is revered as the beef of the north—it's a valuable commodity and part of the reason we have driven fifteen hours across most of British Columbia with a rented canoe lashed to our roof. It's late September, the peak of hunting season in the north, and a prime destination for many hunters is the same place we're headed: the Peace River valley, about 1,100 kilometres northeast of Vancouver, where an estimated 20 percent of British Columbia's entire moose kill takes place.

Most British Columbians do not know that their province stretches east of the Rocky Mountains and includes a mighty chunk of North America's great interior plain. Long after most of the continent had been carved up, its best agricultural land settled, the Peace remained remote and largely unpopulated. Despite its great natural wealth—coal, oil and gas, plus hydro and vast tracts of arable farmland—surveyors spilled into the Peace region only a couple of years before World War I. That sense of pristine remoteness is largely gone now, as this once underdeveloped fringe of the province has been transformed beyond recognition, into what business types like to call an "economic engine."

I'm riding shotgun beside Ben Nelms, twenty-seven, a photojournalist who shoots for Reuters, the Canadian Press, Bloomberg and many other media outlets. He is one of those strange life forms known as a "millennial"—he chews tobacco, doesn't flinch at drinking gin straight out of the bottle and spends entire restaurant meals not looking up from his phone. Ben is a talented photographer as well, a natural as it were, and a steady hand at the wheel.

I'm a freelance journalist, the equivalent in the media ecosystem of a coyote: I scavenge for stories, often gorging on the scraps left behind or missed by mainstream outlets. Over the years I have developed a fascination, a beat of sorts, focused on what the financial markets call "commodities"—things like copper, trees, natural gas and even staple foods (moose futures do not yet exist, to my knowledge). I have always been amazed at how a commodity like electricity—such a prerequisite to our civilization and very survival—is so taken for granted by most of us who consume it. As the enabler of all modern comfort and progress, the process of generating and moving electricity across the continent is, despite its mind-numbing complexity, mostly invisible: with a flick of a switch it appears, as if delivered by some supernatural force. So when I first heard about a big hydroelectric project proposed way up in the northeast corner of BC, out of sight and apparently out of mind, I resolved to know more.

Opposite: Ben documents the wildlife during a rest from paddling in the middle of the Peace River.

Photograph by Christopher Pollon

Ben was a fitting partner in this endeavour. We met on a *Canadian Geographic* assignment in 2012. Over four days we worked together 100 kilometres off the west coast of Vancouver Island on a cramped commercial shark-fishing boat. We were both unprepared for how dangerous the assignment was—no life jackets, no guardrails and giant hooks everywhere. Rogue waves and storms whipped up out of nowhere. We confronted the dangers in silence, but both of us feared the same end: falling overboard at night while the classic rock blared and the crew fished on, our screams unheard as the floodlights receded into the ether.

So with that test under our belts, we resolved to tackle the Peace River valley, which I had taken to calling "the valley of ghosts." That's because the 83-kilometre stretch of river we planned to paddle, between the town of Hudson's Hope and just upstream of Taylor (south of Fort St. John), is the exact section that will be destroyed by BC Hydro's newest mega-hydro project, the recently approved Site C dam. In the next decade, a 60-metre-high mountain of compacted earth will likely stretch more than a kilometre across the main stem of the river, causing the waters behind it to swell into a 93-square-kilometre artificial lake, drowning a stretch of rich farmland. The waters will also swallow fifty islands and a valley that is habitat for farmers, ranchers, trappers and innumerable creatures big and small.

Compared to our shark assignment, this one felt more personal: in our travels across the Peace region we would be little more than tourists, but I was aware that the two existing dams on the Peace supply about a third of BC Hydro's total electricity. Whether we understood it or not, we were already intimately connected to this valley, and that fact made our presence on the river relevant. Important even.

We had come to document a landscape on the verge of an apocalyptic flood.

Many cultures have some version of the mythical world-ending flood—often a creation story where divine punishment rains down upon unsuspecting mortals, courtesy of vengeful or indifferent higher powers. Water is often a means of cleansing, as in the Book of Genesis, when an enraged God wipes the slate clean, employing Noah as righteous enabler and animal wrangler. Hindu legend tells of Matsya, a benevolent talking fish who befriends the world's first man and warns him of a great flood set to destroy the planet. By the time the world has been inundated, Matsya (actually the protector god Vishnu in disguise) balloons into a colossus and tows the man to land, where life can begin anew. Closer to home, the Tsimshian First Nations of northwest British Columbia tell of "water beings" arriving at a potlatch in the form of a flood and thick fog that submerged a great village in the Skeena River estuary; when the water subsided, the survivors established new villages at the points where they were deposited. The world they once knew had been redrawn by the flood.

In coming here we were also inspired by more recent tales, both fact and fiction. In 1968 Edward Abbey published *Desert Solitaire*, which included an account of paddling the doomed section of the Glen Canyon, an irreplaceable stretch of the Colorado River that was flooded to create Lake Powell. "The beavers had to go and build another goddamned dam on the Colorado," the story opens; within three pages Abbey is quoting Shakespeare, Robinson Jeffers and Sir Walter Raleigh.

The premise of our Peace Valley expedition was also the same as that of *Deliverance*, James Dickey's 1970 novel that follows four middle-aged urbanites on a ruinous weekend canoe trip on a fictitious northern Georgia river. Dickey based the story on a real river called the Chattooga, also in northern Georgia, which was dammed to create the Tugalo Dam in 1923. Much of the book's foreboding, sinister atmosphere (*The Lord of the Flies* meets *Heart of Darkness*) stems from the idea that the hostile and unforgiving landscape the men navigate is a sacrifice zone, soon to be erased. A living valley already haunted by ghosts.

The Peace Valley shares little with the lands surrounding the Colorado and Chattooga, other than that it too is being sacrificed in the name of the greater societal good. Or at least that's the official story: Site C was defeated and apparently mothballed in 1983, when an independent regulator determined the project was not in the public interest. More than three decades later it's back, for better or worse.

In coming here, our initial ambition was to ignore the noise surrounding Site C, including the NIMBY versus "jobs at any cost" battle waging in the Peace Country itself, and be agnostic about our assignment. Keep it simple. Float down the doomed section of river valley like nineteenth-century anthropologists and record for posterity what is there. Journalists often take great pains to separate themselves emotionally from a subject and not become a participant—to stand outside, weigh all available facts and somehow be "objective." But we would soon find ourselves floating down the Peace River into an Edgar Allan Poe–sized political maelstrom we had not anticipated. Before this was over, we would have to render judgment. What was the real cost of Site C?

As we discovered, it's a story about wildlife, farmland, natural gas and electricity—and the trade-offs our society collectively makes to live in comfort. It's about the real cost to be paid, both financially and otherwise, to reap the benefits of 5,000 gigawatt hours (GWh) of new energy from the Peace River. And it is about much more: the Peace Valley has been a prosperous home to humans for over 10,000 years. How many stories and lives, human and otherwise, will be erased when the next great flood rises to engulf the Peace River valley?

The Gates on the Peace River, ten kilometres east of Hudson's Hope.

PART ONE:

WATER

DAY 1 | AN UNNATURAL SERENITY

AS I LOOK OUT across the Peace River from Hudson's Hope, the south bank glows crimson and rusty orange. Under a brilliant late September sun, the trembling aspens look as if they're on fire.

Our plan is to spend four days and three nights on the river, paddling about 100 kilometres between here and Taylor (just south of Fort St. John), covering the section of river that will soon be under 50 metres of water. Along the way, we will camp out on the wild, uninhabited islands that dot the braided main stem.

Right: Approaching the Gates, ten kilometres east of Hudson's Hope.

MEC RENTAL

I am not an experienced canoeist—and Ben isn't much better. Before I left, I called Caroline Beam, a teacher who lives on the water just downstream of Hudson's Hope, to inquire about any possible hazards. "You'd have to work at it to get in trouble on that river," she said with a laugh.

A big Chevy pickup truck rumbles up to the boat launch. A young man, not more than nineteen years old, jumps out and assembles a fly rod. Like most of the men who come to the Peace Country to work in oil and gas, he's not from around here. He's from Oshawa, Ontario, Canada's flagging auto-making centre, just east of Toronto. In the past he might have eased onto a General Motors assembly line right out of high school, with reasonable prospects of rising to lower middle class. To do the same now, he has travelled 4,000 kilometres to the boom-and-bust natural gas fields. "The money's amazing," he says without enthusiasm. Is he worried about losing this fishing spot when the new dam is built? "It might make the fishing even better here." The kid clearly feels no connection to this place.

A young man from Oshawa, Ontario, fly-fishes on the Peace River at Hudson's Hope at the spot where we launched our paddling trip.

The Peace River is about 200 metres wide at our put-in, flowing a murky jade green. Our paddling is assisted by a steady current pushing this water eastward across the Alberta border, where it makes an abrupt northward cut to Lake Athabasca and Great Slave Lake. From there it joins the great Mackenzie's main stem, which draws the runoff from about 20 percent of Canada's land mass into the Beaufort Sea of the Arctic Ocean, not far west of Tuktoyaktuk.

The river is animated by circular eddies, which look like jellyfish hovering just beneath the surface. At random points the river boils, belching knuckle-sized bubbles of gas. Large standing waves form out of nowhere, forcing us to manoeuvre our bow to face the wake or risk tipping our overloaded canoe. But challenging whitewater this is not: the Ne-Parle-Pas ("Don't Talk") Rapids—named by voyageurs because the rapids were silent and thus came without warning—that used to surprise and capsize paddlers close to here no longer exist, nor does the great Peace Canyon, which brought mortal fear to explorer Alexander Mackenzie and his team of voyageurs (and their unnamed pet dog) when they attempted its first canoe passage in 1793.

A section of the once-treacherous Ne-Parle-Pas Rapids in 1929, near Hudson's Hope.
Image I-33315 courtesy of the Royal BC Museum and Archives

An unnatural serenity has tamed this stretch of water. That's because beginning in the early 1960s, the Peace Canyon was throttled by 90 million tonnes of earth, rock and concrete, to make way for the W.A.C. Bennett Dam, which, in conjunction with a smaller dam downstream, now generates a huge proportion of BC Hydro's total energy and capacity.

The Peace is the only river on the continent that breaks through the Rocky Mountains, and in doing so, it develops enormous energy as it drops across the Rocky Mountain Trench onto boreal plains on its way to the Mackenzie River. It is this topography that continues to attract hydro developers to the Peace River valley.

By the time the W.A.C. Bennett Dam was completed in 1968, it had indelibly altered not just the river, but the entire region. As one of the largest earth-fill (made from compacted earth) dams ever built at that point, it was a product of its time—reminiscent of the outlandishly bold development schemes of the Cold War era, like atomic scientist Edward Teller's 1950s plan to create deep sea ports off Alaska by detonating hydrogen bombs, or the so-called North American Water and Power Alliance, which proposed to transport water from Alaska and northern Canada to the Lower 48 and Mexico via a continent-long, engineered trough. Underpinning such visions was the conviction that humanity possessed a God-given right to bend nature into any shape required to meet the demands of progress. Indeed, the slogan of the US Bureau of Reclamation, the body responsible for transforming the Columbia River beginning in the 1930s, said it all: "Our Rivers: Total Use for Greater Wealth."

The Peace shares much with that great river to the south. In *A River Lost: The Life and Death of the Columbia*, Blaine Harden recounts how the Columbia was transformed from the world's greatest salmon river into a "machine river": a series of slack-water puddles controlled to spill water remotely by push button. With this destruction came trade-offs that are impossible to deny:

> Dams gave people who lived in the Pacific Northwest the cheapest electricity in the country. They turned the deserts of eastern Washington and Oregon into gardens. Their power made aluminium for the airplanes and fuel for the atomic bombs that helped win World War II. The sum total of the concrete miracles, over half a century, transformed Washington, Oregon, Idaho, western Montana, and the Canadian province of British Columbia from a boondock into a high-tech, high-wage region whose gross national product ranked tenth in the world.

Today the power generated by both the Peace and the Columbia feeds into the same colossal Western Interconnection—one of three principal electric networks in North America, and one of the technical wonders of the world. This grid's tentacles reach across half of the total land mass of the United States and into Canada, bringing civilization to 50 million people from northern British Columbia to the Texas border, and as far east as Nebraska. (The so-called Eastern Interconnection is even more vast, enabling power to flow from Ontario to Florida.)

Washington State had senators Henry M. Jackson and Warren G. Magnuson to pork barrel the biggest of the Columbia dams into existence. British Columbia and the Peace had its own advocate in Premier W.A.C. ("Wacky") Bennett, who led BC from 1952 to 1972. In 1952, Bennett led to power the Social Credit Party, a party popularized by Alberta preacher "Bible" Bill Aberhart—under whose premiership the movement spread through Depression-era Alberta like a brush fire. The newly elected Socred premier of BC would go on to champion what became known as the "Two Rivers" strategy, a vision to harness the Peace River and the BC section of the upper Columbia River to fuel the industrial development of the entire province.

The seed to develop the hydro resources of the Peace had been planted by a Swedish industrialist-tycoon named Axel Wenner-Gren in the years following World War II. Wenner-Gren made his vast fortune selling vacuum cleaners, shifted to arms dealing, and became even richer from selling weapons to both sides during the war. But by 1957, his complicity with the Germans was apparently forgiven. In that year he unveiled a grandiose scheme to "open up the BC north"—including construction of an ultra-futuristic monorail system, hydro dams and pulp mills. He planned, with the blessing of the BC government, to embark on a $5-million survey of all the timber, mineral and water-power resources across more than 10 percent of the province, including much of the Rocky Mountain Trench. Public outrage ultimately killed the plan (with critics calling the province "Swedish Columbia"), but the survey work was one of the first serious examinations of the hydro potential of the Peace River. In a few years Wenner-Gren was gone, but the newly nationalized BC Electric Company, reborn as BC Hydro under W.A.C. Bennett's guidance in 1962, took the vision to harness the Peace Canyon for hydro and ran with it. "The Peace is an empire crying out for development," Bennett told a scrum of reporters in 1957. "Its hydro, coal, gas and oil make it the greatest potential energy resource area in North America."

W.A.C. Bennett at the official opening ceremony for the dam that would take his name, September 22, 1968.
Hudson's Hope Museum and Archives, HHMA1981.BEA.007

Beginning in the early sixties, as construction began on the W.A.C. Bennett Dam, the main stem of the Peace River was blocked and a reservoir ballooned in size behind the walls, eventually holding the drainage of an artificial watershed bigger than Ireland. This ended forever the seasonal freshet that would

Above: A conveyer dump at the bottom of W.A.C. Bennet Dam during early days of construction.
Hudson's Hope Museum and Archives, HHMA1981.Ben.006

Above right: Looking east over the powerhouse, June 13,1967.
Hudson's Hope Museum and Archives, HHMA1990.002.001

recharge and nourish the valley with spring floodwaters. Warmer water from the reservoir meant the reach between the dam and Taylor—about 100 kilometres downstream—would never again freeze completely over.

Behind the 60-storey-high (183-metre) dam walls is the Williston Reservoir, BC's largest freshwater body, which engulfed over 1,770 square kilometres of land, lake and river. In its place was a four-armed man-made lake so colossal that Bennett famously bragged it would change the climate of the north. (Local farmers today claim it did.) Damming the Peace flooded enormous sections of the great Finlay and Parsnip river systems. Before the dam, the wild Finlay roared down the Rocky Mountain Trench from the northwest, joining the Parsnip River at Finlay Forks, where the combined force of water turned eastward, giving birth to the Peace River. To this day the true scale and extent of the destruction of life in these drainages can only be guessed at—no baseline information was recorded at the time to accurately measure what lived there before the great flood.

Just downstream of that is the much smaller Peace Canyon Dam, which was built in the late 1970s and began operating in 1980. Site C is the third dam originally envisioned by Bennett and BC Hydro on the Peace River—a project that was born in the

1950s, rejected repeatedly in the 1980s and early 1990s, but which remained on the books in a state of perpetual, zombie-like readiness, ready to lurch back to life given the right conditions. It was approved by the government of Premier Christy Clark in December 2014.

Today W.A.C. Bennett remains a potent influence, popular with Clark's booster-ish, pro-development government. In September 2015, as multiple court cases from local landowners and First Nations aimed at halting the project's early construction lingered in appeals, Clark appointed W.A.C. Bennett's grandson Brad Bennett as chairman of the board of BC Hydro, the Crown corporation that—in addition to building both existing Peace River dams—is now pushing forward with Site C.

In autumn 2015, just as we are making our first tentative strokes on the river, the first construction is beginning in earnest. The only thing new about the plans for Site C is the upwardly revised $8.8 billion price tag (making it the most expensive infrastructure project in BC government history).

The confusing thing about Blaine Harden's description of a machine river is that the words do not match what we experience on the water. If a river is dismissed as a spent force, damaged goods so to speak, it might be seen as something that is already so diminished that another dam will make no difference. But this is not the case: the Peace River remains a nursery of extraordinary abundance. The valley and vicinity make up a biodiversity sweet spot, where the foothills meet spruce forest and mixed-wood ecosystems, with a mingling of plants and animals from each. It is home to about 60 mammal species, at least 215 kinds of birds, 6 types of amphibians, and about 30 species of fish. The native bison that once roamed here were gone by 1906, but one of the continent's last free-roaming bison herds (which gained a foothold in the Peace after about fifty imported plains bison escaped captivity in the 1970s) is found just to the north around Pink Mountain, in the Halfway and Sikanni river valleys.

The abundance of the Peace Valley speaks not only to the resilience of the river and the life it supports, but also to the wealth of life that must have existed before the dams were built.

Just past the point where tiny Lynx Creek drains into the Peace, the air becomes so thick with emerging mayflies that I feel as if I'm paddling through a snowstorm. I open my mouth to exclaim in amazement to Ben, only to swallow two hatchlings like a giant trout.

Right: Newly hatched mayflies rise to the surface of the Peace River, southwest of the Halfway River confluence.

Below: Bonaparte's gulls feast on insects emerging on the Peace River, a few kilometres west of the Halfway River.

From the low vantage point of our canoe, it's apparent that river-born insects form the basis of the great abundance. The humble midge (which looks like a mosquito when mature), mayflies, caddisflies and stoneflies are at any moment in the process of procreating, emerging and being devoured en masse. Most of the fish along this stretch have very specific dietary preferences: mountain whitefish prefer caddisflies, the most abundant aquatic bugs found here; arctic grayling love mayflies; and longnose suckers favour midges. Rainbow trout will inhale just about anything they can find.

During this first day of paddling we pass a great variety of islands, most of them provincially owned Crown land. There are more than fifty such islands in this reach of the Peace River—uninhabited by humans, but otherwise heavily populated. The island we choose to camp on for the night is long and shaped like a flying saucer, bulging and gaining elevation in the middle. The island fringes are dominated by yellow and rust-coloured poplar and paper birch, while the plateau is sheltered by thick spruce.

Our first campsite, on an island about 15 kilometres east of Hudson's Hope. Such islands provide a critical protective buffer zone for ungulates against bears, wolves and other predators.

Islands like this provide a critical protective buffer zone for ungulates such as elk, mule deer, white-tailed deer and moose, beginning in the spring and continuing into the summer. Many newborns will spend their first month of life in such places, at a time when they are most vulnerable to predation. Islands are relative safe havens because even though all the predators are good swimmers, most will choose easier pickings on the mainland.

Abundant moisture is the secret to what makes the islands (and riverfront land) so unique and irreplaceable. Proximity to water means that balsam poplar and birch grow much larger than similar trees on the drier, surrounding plateau. Great cottonwoods with bark like plate armour line the riverbank and island fringes, providing prime nesting sites for eagles. These large-diameter trees become maternal birthing

The light from Hudson's Hope glows across the hills looking southwest from our first campsite.

I unwind by the fire during our first night on the river. There was no shortage of firewood on this densely forested island set on the main stem of the river.

We paddled into this small cove, which sheltered our canoe from the flow of the river, on our first night.

dens for ungulates, fishers (weasel-like predators with faces like mini Ewoks) and the multiple bat species that live in this valley. Moose rely on this riparian zone too, particularly during heavy winter snowfalls. Here, warm coastal air penetrating the Rockies means less snow and improved mobility in evading predators. Moose also gorge on the red alder and willow that thrive in this wet environment.

We make for the high ground of the island plateau to set up camp: it's a sanctuary protected by a wall of dense conifers. It's our own private island set in the main stem of the Peace River, and we are completely alone. Or almost. Hugging the lower-lying sections of the north shore of the river is Highway 29—a relatively recent development in this part of the world. Hudson's Hope was connected by road to Fort St. John when the Halfway River Bridge was built in 1938, with upgrades to facilitate the hauling of coal in the late 1940s. As big tractor trailers rise and fall on the hilly road, the distance and speed bends the roar of their air brakes. Many of the trucks are transporting wastewater to Fort St. John from nearby gas wells in the great Montney shale formation, of which this valley is a part.

I set off to explore our island with the fading daylight. There are animal tracks

Above: Bald eagles are especially fond of the large-diameter cottonwoods that grow within the Peace River riparian zone. BC Hydro plans to mitigate the destruction of nesting sites by erecting wooden poles along the new Site C reservoir banks.
Thinkstock/iStock/S_Lew

Right: The province estimates there are about fifteen thousand grizzly bears like this one in BC today. No one knows how Site C will affect grizzly populations in the Peace Valley.
All Canada Photos/Roberta Olenick

everywhere. The trails have been packed hard by island-hopping ungulates. It's getting too dark to see, but as I approach our camp, there are enormous paw prints. Grizzly or black bear I have no idea. I'm no ranger, but they seem fresh.

We were warned just before we left that a grizzly sow with cubs had been sighted on an island just upstream of this spot. Weeks before, national media covered the story of two hunters near Fort Nelson (about 400 kilometres north of here), who surprised a similar grizzly with cubs while hunting. The men climbed over a ridge and came upon the bears. What followed was a classic defensive attack: a lightning-fast mauling with claws and teeth focused on the heads of the victims, followed by a quick retreat. Each suffered serious injuries to the head, neck and arms, but survived the attack, which lasted a total of thirty seconds. Understanding that the bear's motive was to protect its young, conservation officers did not attempt to track or shoot her.

Grizzly bears are known to move through this valley, including the nearby north–south river drainages at Farrell and Cache Creeks, but no one, including BC Hydro, knows much about their numbers or movements here. Unlike black bears, which have never naturally existed outside North America, brown bears in general, collectively known as *Ursus arctos*, are the most widely distributed bear species on earth. But they are currently either extirpated or declining across much of their original range.[1] A number of things put grizzlies at a disadvantage to black bears: females not only reach reproductive maturity later, but they also have smaller litters. Grizzlies are also more likely to get into conflict with humans, because they are by nature more aggressive and less intimidated by people. Exacerbating matters closer to home, grizzly hunting continues to be a trophy sport in British Columbia. The provincial government estimates there are about 15,000 grizzlies today, and an average of 300 are legally killed by hunters each year.

Upon reflection, it's absurd to be worried about bear attacks on this tiny island, especially given what's coming to this valley over the next decade. Bears in these parts have much more to fear from humans.

Sleep on this night is uneasy; at regular intervals I hear small landslides from the canyon walls on the south bank. It sounds like sand falling through a giant strainer. Locals call it sloughing (pronounced "sloffing"), which is the sliding of the clay, shale, sand and silt from the cliffs that rise on both banks of much of the Peace. It's one of the biggest concerns locals have about the Site C project moving forward.

A jet boat heads east toward Hudson's Hope against a cliff backdrop of eroding silt, clay and sand.

To understand the threat of sloughing, you need to go all the way back to the great thaw, which followed the last ice age. The area was once at the bottom of a vast post-glacial lake (Glacial Lake Peace), created by dams of ice as the last continental ice sheets melted more than ten thousand years ago. Over time Glacial Lake Peace laid down layers of sediment and silt (on top of much older layers) that can be seen (and heard) along the canyon walls that line the river. When exposed to water, these graded layers of silt, clay and sand can become like a sticky soup, which means that the cliffs we see all along the Peace will be prone to collapse and erosion when the Site C reservoir rises against the canyon walls.

Much of the region is undergirded by these same ancient sediments, which is just

The fine silt, clay and sand banks that line the river are constantly eroding. When the flood comes, much of this material will collapse.

one reason why the Williston Reservoir at the W.A.C. Bennett Dam ended up being so controversial when it was created in the 1960s. It was there, as the flood waters rose, that the soft sedimentary banks of the newly flooded reservoir began to erode—abetted by huge shore-lapping waves. In the mid-1980s, twenty years after the lake was created, the banks had still not yet stabilized.

It wasn't supposed to be this way. In *This Was Our Valley*, the most thorough account of the flooding that created the Williston Reservoir, Shirlee Smith Matheson (with pioneer/poet Earl K. Pollon, no relation to author) notes that BC Hydro initially predicted the new reservoir would enable the development of new forest and mineral resources, and provide access to fish and game once denied to "all but the most

Opposite top left: A family at Fort Grahame, on the banks of the Finlay River, 1940.
Image H-04671 courtesy of the Royal BC Museum and Archives

Top right: Sekani chief Charlie Hunter and family at Fort Grahame, June 6, 1914, as photographed by Frank Swannell.
Image I-33182 courtesy of the Royal BC Museum and Archives

Bottom: Fort Grahame on the Finlay River, as photographed by Frank Swannell in 1939, less than three decades before it disappeared beneath the Williston Reservoir.
Image I-33492 courtesy of the Royal BC Museum and Archives

ambitious sportsmen." Hudson's Hope, long sidelined from the coal and forestry businesses by its remote location, would finally become Canyon City, a centre of industry and commerce.

But the water rose faster than BC Hydro anticipated.

According to Tom Roberts, a former logger and Peace resident quoted by Matheson, very little prelogging of the area to be flooded was ever done, largely because it was not considered practical to build the roads necessary to truck it all out. For the areas that were logged, the rising floodwaters drowned out many existing logging roads, submerging harvested logs before they could be taken out. The end result was the creation of a vast aquatic wasteland so thick with woody debris that in places one could walk across the surface. Despite Hydro's salvage efforts, Roberts estimated there was 34,000 acres' worth of woody debris floating on the reservoir by the time it was filled. The wood that the small army of loggers did manage to remove in time—including many trees more than a century old—was given away to forestry companies for virtually nothing.

Matheson recounts how bush pilot Pen Powell, who owned and operated a nearby hunting and fishing lodge, counted more than a hundred moose trapped on a high piece of ground by floating woody debris as the flood came. The next day he flew by to find the land gone, with floating moose corpses dotting the surface. An estimated 12,500 moose were killed by the flooding.

By far the worst fate was reserved for the First Nations of the Rocky Mountain Trench, historically and collectively known as the "Sekani," whose pristine homeland, centred around the Finlay River watershed, was destroyed by the dam. Today the survivors of the Williston Reservoir flooding include residents of two Rocky Mountain Trench First Nations whose communities include: Tsay Keh Dene, home to members of the Tsay Keh Dene Nation, which is situated at the north end of the Williston Reservoir, and Fort Ware, home to the Kwadacha First Nation, about 70 kilometres northwest of the tip of the Williston Reservoir.

Matheson recounts how BC Hydro sent Charlie B. Cunningham, a former big game hunter and guide, on a 1962 expedition up the Finlay into the homeland of the Sekani, who once hunted, trapped and fished throughout north-central BC. His mission was to warn them about the coming flood. He marvelled at the prosperous settlements built in bucolic settings at Ingenika and Fort Grahame, whose inhabitants

Sikanni Chief Charlie Hunter
Fort Grahame
June 6/14

are today known as the Tsay Keh Dene people. (The Finlay, the biggest and mightiest tributary of the Peace, ran so pure that everyone drank from the river.) The people he encountered were still deeply immersed in the rhythms of a hunter-gatherer lifestyle: starting in spring each year, they followed the seasonal harvests of fish, moose and berries.

"So far I have not found anyone hostile to the flooding," Cunningham wrote of the visit. But to most of the people he met, the only dams they had ever known were built by beavers, a historical staple of their traplines. When the great flood finally came, it was as if the end of the world had arrived. Fort Grahame's cemetery was flooded, while Ingenika Point's graveyard was moved inland by BC Hydro to escape the floodwaters. It wasn't moved far enough, though, and the recently exhumed remains eroded into Williston Reservoir.

Attempts were made to resettle the Fort Grahame band at two locations—one was never inhabited and another was only inhabited for a short time before the people tried to move back north into their territory. One of the sites the people moved back to was Ingenika Point at the northern tip of Williston Reservoir, which was located close to the mouth of the Ingenika River. But here their access to the upper Finlay was prevented by a debris trap infamously known as "the Plug"—an eight-square-kilometre log jam of huge cottonwood, spruce and poplar all pressed together on the reservoir. No watercraft could penetrate it. Shortly after the band's resettlement, logging companies descended on their new home and razed the forests made newly accessible by the reservoir. The place was now worse than a wasteland. When the dam lowered the lake level by nine metres late that first winter, it exposed vast tracts of silty lake bottom to powerful winds, generating dust clouds so thick that the settlement was invisible to passing planes. (At Tsay Keh Dene, dust problems persist to this day and are a major health concern for the people.) The local men who dared to launch their boats into the new reservoir often became stuck for days, stranded in treacherous aquatic log jams, at the mercy of the winds that stirred up the surface.

In 2006 Garrett Seymour, a seventeen-year-old senior at the Aatse Davie School in Fort Ware, created a spoken word diary about life after the flood. "In my grandfather's time, the river was a road," he says as a camera follows him on a walk through his village. "The rivers were changed forever in 1968. That was a long time before I was born, but the effects of those changes have deeply affected how my people live today. Some people thought the end of the world was coming when the waters came up. Some families never recovered from the shock. The environmental trauma to my land is very great. The trauma to my people is equally as great."

A lasting irony of this little-known episode in BC history is that nearly fifty years after the First Nations of the area lost everything so the W.A.C. Bennett Dam could be built, both the Tsay Keh Dene Nation and Kwadacha First Nation, are still forced to live off the BC electrical grid. They subsist on dirty, expensive and unreliable diesel generation for electricity. (The only power line running through Tsay Keh Dene

An abandoned cabin on the edge of the eroding Williston Reservoir in 2008.
Photograph courtesy of PVEA Archives

territory today was built to connect to a mine site near Thutade Lake, the headwaters of the Peace and Mackenzie Rivers.)

"BC Hydro provides reliable power, low cost for generations, they say in their brochures," says video narrator Seymour, who still lives in Fort Ware, dividing his time between hip hop music and seasonal mining work. "Sometimes in our community they have power failures every day for weeks. The cost is high. My dad [who was handicapped in a trucking accident on a logging road] can tell you of his incredible power bills to operate his chair."

In June 2016, when a new First Nations exhibit called "Our Story, Our Voice" opened at the W.A.C. Bennett Dam visitor centre, the Tsay Keh Dene Nation did not participate. An inscription on the wall of the exhibit quotes Chief Dennis Izony, who says the nation's absence was "due to the on-going trauma and lasting effects of the creation of the reservoir on our nation and its people that has yet to be resolved."[2]

Meanwhile, the Williston Reservoir is today not the recreational amenity envisioned so long ago. "Williston Reservoir is a very large and potentially hazardous reservoir for the unwary," reads a BC Hydro community recreation web page for

A dust storm in 2008 kicks up eroded sand and silt from the exposed Williston Lake bottom during a drawdown of lake levels.
Photograph courtesy of PVEA Archives

Williston Lake, which warns boaters about the myriad dangers, including floating debris and unstable banks. The site goes on to warn snowmobilers to exercise "extreme caution" due to pressure ridges, gas holes, open water and broken shoreline ice.

So it's understandable if some locals are hesitant to celebrate at the suggestion that the flooding of the remaining section of valley for Site C could drive tourism and even perhaps a future real estate boom.

Woody debris washed up on the shoreline of Williston Reservoir in 2003. Forest waste from an estimated 34,000 acres of land was floating on the reservoir by the time the W.A.C. Bennett Dam was completed in the late sixties.
Photograph courtesy of PVEA Archives

Our first night on the Peace, spent about 40 kilometres upstream of the Williston Reservoir, is mostly uneventful. As I lie in my sleeping bag, I replay the history of how the W.A.C. Bennett Dam changed the Peace region. I am new to this part of the world, and these revelations have caught me off guard. I always presupposed that the construction of this mega-dam in particular was an act of supreme foresight on the part of past leaders, a gift to future generations, and a major reason why BC emerged as a prosperous resource powerhouse in the first place. All of which rings true to a degree, but with this history lesson came a dawning recognition that the energy sources underscoring our collective post-war prosperity have come at a massive, little-known cost to the Peace region.

At some point later in the night, we are jolted awake by horn blasts from a tractor trailer, but the sound is not coming from the distant highway. We don't know it yet, but we have heard the mating call of a bull elk—which can stand 1.5 metres tall and weigh 330 kilograms—and is the most common ungulate species on these islands. During this time of year, such a beast feeds so voraciously that it can pack on a kilogram of weight every two days—necessary for the energy demands of maintaining a harem of up to fifty females.

"Even asleep, you are a coiled spring," writes Bill Bryson of the urbanite experience of sleeping outside in the wild. I'm a bit wound up myself, but nothing short of a bear claw ventilating the tent will extract me from my sleeping bag on this night. The temperature has dropped to zero, but the bag is cosy down to at least minus five.

DAY 2 | OUR MACKENZIE MOMENT

WE WAKE TO the smouldering embers of last night's fire. Coffee is priority one, followed by a hike to the island's shoreline to catch our first fish.

Despite the impacts of two dams, the Peace River today is home to nearly thirty species of fish, including rainbow trout, northern pike and two species of whitefish. The Arctic grayling, now endangered in the watershed of the Williston Reservoir, is a fish with a stunning blue and mauve sail-like dorsal fin. And the fat-rich bull trout, a sport fish prized for eating, is currently on the provincial blue list, meaning it is at risk across much of British Columbia.

We stopped to fish at many points during our journey, catching multiple bull trout and two species of whitefish.

Just as I retrieve my first cast to shoreline, a bull trout pounces on the lure. The char is no longer than my hand and is a thing of beauty—delicate spots of yellow darken progressively into orange up its streamlined, bullet-shaped body, set against a backdrop of gun-metal grey that darkens to black. Bull trout are indigenous to these waters—voracious aquatic predators with an insatiable appetite for bugs in their juvenile years, then moving on to fish as they grow.

It's strictly catch and release for bull trout in September, but I probably wouldn't eat this fish even if I could. That's because the flooding of the Williston Reservoir caused an unnatural spike in methyl mercury in much of the Peace River system.

High in fat and prized for eating, bull trout are native to the Peace River and are currently on the provincial blue list, meaning they are at risk across much of BC.

Organic matter decays and naturally deposits inorganic mercury into soil, but when vast areas are flooded, sulphate-reducing bacteria can transform this inert inorganic mercury into toxic methyl mercury. This metal concentrates in the bodies of predators, in greater amounts the higher the animal resides on the food chain.

West Moberly First Nations chief Roland Willson says, "Bull trout are our salmon," and herein lies a problem. Willson and his family rely on these fish as a part of their regular diet, occupying seasonal fish camps on the Crooked River (now a Williston Reservoir tributary) to target the fish, which migrate in spring from the reservoir into many smaller rivers and streams. In early 2015, the West Moberly and McLeod Lake First Nations released a study that analyzed about sixty bull trout taken from the Crooked River. Results suggested that a steady diet of local bull trout could be exposing band members to levels of mercury that were dangerous, particularly for children and pregnant women.

The Crooked River findings made a short splash in Canadian media, when Willson delivered 90 kilograms of mercury-tainted fish to the BC Legislature in Victoria.

"We can't eat it, so I'm offering it back," he said on May 15, 2015, from the lawn out front. For years the government's own regularly published Freshwater Fishing Regulations have warned anglers that methyl mercury levels in Williston Lake bull trout (and lake trout) "may be high," and that "normal consumption is not a significant hazard to human health, but high consumption may be." Willson told me later that this warning is intended for occasional fish eaters only. "First Nations consumption is typically much higher. No studies have ever been done to see what the consumption levels are with First Nations." He remains worried that the same methyl mercury issues that emerged after the previous reservoirs were created will affect the fish on the Moberly and Peace Rivers after the Site C dam is constructed. (West Moberly is one of about seven First Nations whose reserve lands form a wide arc surrounding the stretch of Peace River that will be flooded by Site C.)

West Moberly First Nations chief Roland Willson hoists a bag of mercury-tainted fish from the Williston Lake watershed, on the lawn of the provincial legislature in May 2015.
The Canadian Press/Dirk Meissner

As for this stretch of river we are paddling, BC Hydro acknowledges that—after construction of the Site C dam—there "may be" an increase in methyl mercury concentrations in fish in the reservoir for a period of time, before amounts return to pre-project levels. Meanwhile four of the fifty recommendations made by the Joint Review Panel struck by federal and provincial governments in August 2013, to weigh the effects of the project and gauge their significance, concerned the methyl mercury issue, including the need to ensure local fish are monitored for mercury if the project moves forward.

West Moberly has emerged as the most outspoken First Nation opponent of Site C, a reflection of the direct impact community members will feel when the lower Moberly River is inundated not far from their reserve. More than 11 kilometres of the lower Moberly will be flooded out, becoming part of the reservoir. Like the other First Nations in the immediate area, West Moberly's reserve is tiny—just over 2,000 hectares in total. But this small reserve size is misleading. As a signatory to Treaty 8, West Moberly has a treaty area that encompasses a much wider 840,000-square-kilometre chunk of northeast BC, Alberta, Saskatchewan and the Northwest Territories. (Other than the Nisga'a Treaty of 1998, Treaty 8 is the only major treaty with BC First Nations ever signed by the federal government.) When it comes to harvesting species like bull trout, even though the reserves are tiny, a signatory to Treaty 8 has the right to hunt and fish throughout the entire treaty area, which includes game and fish-rich sections of the Peace that will be destroyed.

For the many First Nations that eventually signed Treaty 8—including eight BC First Nations from the Sekani, Slavey, Beaver (Dunne-za), Cree and Saulteau ethnolinguistic groups—the agreement originally addressed the fallout of the Klondike Gold Rush. Gold fever saw tens of thousands of prospectors spill into the north, a migration that peaked the year the treaty was signed. Prospectors flooded across the banks of the local Halfway and Finlay Rivers in 1898 on their way to the Klondike, while previous smaller gold rushes—beginning in 1862 on the Peace and Finlay Rivers,[3] then on the Omineca in 1870—also brought brief but world-altering migrations into the region. The influx of prospectors effectively ended the fur monopoly of the Hudson's Bay Company in the Peace, because failed miners turned in droves to trapping (and anything else they might try) in order to survive.

It was this rapacious Klondike-bound migration in particular that spurred the federal government to impose Canadian sovereignty over much of the sparsely populated hinterland. The problem, of course, was that First Nations had been there for millennia and relied on the land for all aspects of culture, life and survival. Records from the first Mountie patrols through the Peace River Country and the north, beginning in 1897, provide a record not only of growing First Nations alarm over the intrusion of outsiders onto their hunting grounds, but also of the tension stemming from the federal government's first attempts to enforce Canadian laws, including restrictions on hunting.

Treaty 8 dates back to 1899, and it is clear now that it meant very different things to its different signatories. Beyond demonstrating sovereignty to unruly American prospectors, the federal government sought to brush aside aboriginal claims to the land in order to tap the great resources of the north. To the First Nations, it was a friendship pact to ensure that non-Natives did not degrade their land, culture and food resources. On September 17, 2010, four Treaty 8 First Nations, including West Moberly, signed a declaration rejecting the Site C dam, with the support of twelve additional First Nations.[4]

I asked Matthew Nefstead, a lawyer with Devlin Gailus Westaway (the firm working on Site C litigation on behalf of the West Moberly and Prophet River First Nations), about how Treaty 8 relates to the dam and the wider issue of resource extraction in the Peace. "The key point is that the Treaty promised that the same means of earning a livelihood would continue after the Treaty as existed before it," he said. "And that they would be as free to hunt and fish after the Treaty as they would be if they never entered into it; and that the Treaty would not lead to 'forced interference with their mode of life.'"

Central to this mode of life is the right to access food fish, and BC Hydro has conceded that Site C will cause "significant adverse effects on fish and fish habitat." With the shift from river to deep reservoir ecosystem, the biggest winners will likely be species such as kokanee and lake trout, while lake whitefish, burbot and rainbow trout are presumed to benefit as well. For the fish more closely adapted to river life—such as Arctic grayling that rely on the Moberly, or the fickle bull trout that migrate up to 100 kilometres to find the perfect spawning habitat on the Halfway River—the future is dim.

In such dire instances "mitigation" becomes a magic word to BC Hydro: it is the company's way to compensate for all the things that will be irrevocably lost. Take for example Hydro's plan to mitigate the destruction of eagle nesting sites, built mostly atop old riparian cottonwoods, by erecting wooden poles near the banks of the new reservoir. For threatened bull trout, Hydro has proposed to mitigate impacts by building a fish ladder at the foot of the dam. It will be designed to enable fish to ascend ten metres to a "trapping pool," where they can be counted, collected and driven in trucks upstream of the dam. One BC Hydro plan to build a "trap and haul" and fish ladder solution is estimated to cost $25.5 million, with annual operating costs of about $1.5 million, presumably to last over the life of the dam.[5] Regardless of how the impacts are mitigated, a 2014 technical review by the federal Department of Fisheries and Oceans states that in spite of the new reservoir being larger in size and volume, the new environment will be a vastly diminished place in terms of biodiversity: "It will likely be unproductive due to the nature of the inputs from the upstream reservoirs, the high flushing rate, shoreline instability and the potential for turbidity in the eastern half."

Back on the river, paddling feels easier today. We have learned to "ride the express" as I call it—manoeuvring the canoe close and parallel to the riverbank, where the current flows faster and does more of the work. Our destination today is the Halfway River, about 30 kilometres downstream. It's at this point we expect to see most of the human activity on the river, mostly from sport fishermen targeting bull trout. Like the Peace River itself, the lower reaches of such tributaries will balloon with the flood—BC Hydro estimates over 15 kilometres of the lower Halfway River will be flooded—forming the club arms of an unnaturally deep reservoir. We soon

The north side of the Peace River approximately ten kilometres east of Hudson's Hope, just before we paddled through the Gates.

approach "the Gates": two spectacular stone bluffs that frame the river. It's here that we stop paddling altogether for a spell, letting the current drive us forward as we cast our fishing lines. Ben's wet flies whizz uncomfortably close to my ear.

The Peace is what river experts call a cobble-gravel bedded river, meaning the bottom is composed primarily of large stones. Before the dams, the annual spring freshet came with such force that these rocks were in constant motion, gouging and reshaping the riverbed as they moved. The great seasonal floods made this a dynamic place. Now that the freshet has been tamed by the dams, the Peace River has simply shrunk inside its former channel margins. Without the great annual rearrangement, the sand and gravel bars have taken on a permanence that is alien to the flood plain this once was.

At one point great masses of petite, elegant Bonaparte's gulls descend upon the

Bonaparte's gulls inhale insects from the river's surface.

water, inhaling surface hatches of insects. Brown-headed, serrated-billed mergansers streak by the north bank, their bodies almost touching the water. And the cacophony of Canada geese is so common here, it has become the background soundtrack of our days on the river.

Winged critters truly abound here. BC Hydro consultants have reported that at least twenty-five active bald eagle nests have been documented in the area, with eleven within the potential Site C reservoir. (Locals say the numbers fluctuate widely from year to year.) Years of surveys beginning in 2008 found 116 songbird species, while 60 species of waterfowl and water-associated birds, including 7 red or blue-listed species, were detected between 2005 and 2008. There are 8 species of owl here, including great horned, short-eared and boreal owls. Broad-winged hawk, northern goshawk and northern harrier also hunt along this reach. And 6 bat species have been confirmed to be "present and reproducing" in the Peace River corridor.

For birds and many other migratory animals, the valley is a component of the "Peace River break"—a critical section of what conservationists call the "Yellowstone to Yukon" (Y2Y) wildlife corridor. The corridor forms an ecologically intact north–south pathway along the Rocky Mountains and surrounding lands on the route from the Arctic Circle to Wyoming. And it is the Peace River break—an area that includes the Williston Reservoir and most of the river covered by our paddle trip—that forms the thinnest section of the entire international corridor.

The idea behind Y2Y is that parks and protected areas are little more than green blobs on a map. Beyond these borders, wide-ranging migratory animals often cannot survive. In addition to birds, this group includes grizzly bears, grey wolves, wolverine, caribou and lynx—animals that need large unfragmented areas to roam. Already subject to enormous development pressures, this corridor will be further degraded by the creation of the Site C reservoir, which will impact the migratory pathways through the bottom lands of the valley. Unstable banks created by the Site C flood will restrict movement for many land animals that routinely migrate across the river. (Virtually all the mammals here are good swimmers.) A case in point is the Halfway River and its north–south valley, which is a vital transit path for ungulates (particularly moose) and all the predators that follow after them.

Just downstream of the confluence of the Halfway and Peace Rivers, in the protected shallows of a long island hugging the mainland, Ben and I encounter an

astonishing sight. At twilight, in the midst of a thick hatch of bugs, I count at least twenty fish breaking the surface at any given second. It goes on for minutes.

All my life I have read and listened to forlorn tales, tinged with awe and regret, of the loss of areas of great natural abundance in Canada. Many of these stories take on the dubious aspect of legend: Vancouver Island rivers running so thick with coho salmon that pioneers had to walk over their backs to ford a stream, or Newfoundlanders collecting great drafts of cod by simply tossing buckets into the sea. For Ben and me, this is one of those moments. We're conscious, too, that in our awe of this twilight feeding frenzy, we have fallen victim to "shifting baseline" syndrome: what to our eyes in autumn 2015 is wondrous abundance surely must be a sad fraction of what once thrived here.

Rare moments like this make us imagine what sights Alexander Mackenzie, the

I make a quick adjustment to the canoe near Lynx Creek before we paddle the river again.

first European to come through these parts, must have witnessed in the 1790s as he paddled this very stretch of river. His journals, which were bestsellers in his day (Napoleon Bonaparte had a volume translated into French to bring into exile on the isle of Saint Helena), provide some idea:

"Mr. Mackay, and one of the young men, killed two elks, and mortally wounded a buffalo... The country is so crowded with animals as to have the appearance, in some places, of a stall-yard." It is believed that Mackenzie was referring to a spot close to present-day Bear Flats. At the time that journal entry was written, he had recently left present-day Alberta on his way to the Pacific, travelling westward along the Peace River, through the break in the Rockies.[6]

The Peace region at that point was the northern limit for the now-extinct American passenger pigeon, which were once so numerous that their migrations could impose darkness on a clear day for hours on end. As Mackenzie noted, buffalo still roamed the region during his 1793 travels, and he reported seeing them as far west as the Parsnip River—not far from the current location of Mackenzie, BC. The buffalo he saw (and ate) were descendants of a much larger species of bison that first migrated into North America from Asia about 200,000 years ago via the Bering land bridge connecting Siberia to Alaska. Buffalo bones more than 10,000 years old have been found at Charlie Lake Cave just north of Fort St. John, bearing the marks of stone tools handled by the early hunters who feasted at the site. It's possible that these humans, whom archeologists suspect emigrated from the south into the area, lived among woolly mammoths, horses (slightly smaller in size than the modern horses we know), muskoxen and a species of camel resembling a modern dromedary. (Horses and camels were both native North American species—they migrated to Asia via the Bering land bridge.)

By Mackenzie's time, the great herds of buffalo were concentrated from central Alberta into south-central Saskatchewan and down through Montana to Colorado. While these enormous herds did not exist in the Peace, the large quantity of buffalo bones found at Rocky Mountain Fort demonstrate that these animals were an important staple used to provision the early fur trade as it penetrated into what became British Columbia. The last buffalo in the region was reportedly shot dead in 1906 near Fort St. John.

The arrival of Mackenzie, by canoe along the Peace River in 1793, was not a complete surprise to the local First Nations. It is said that the arrival of white men and the fur trade had been foretold in advance by a powerful Beaver (Dunne-za) First Nation prophet, a visionary named Makenunatane. Dreamers of great power were known to the Beaver as swans, because it was said their souls could fly. Swan Chief Makenunatane saw maps in his dreams as real as anything on paper, depicting the converging pathways of animals, humans and the spirit world. *The Prophecy of the Swan*, a book about the early upper Peace River fur trade, reports that the arrival of the white man saw the people look to Makenunatane for direction. He counselled that they should embrace the fur trade and Christianity, and he personally transitioned from a chief of the hunt to a chief of trade.

For his part, Mackenzie would go on to brave the Peace Valley canyon—where, facing a near mutiny from his crew, he was forced to portage around it. He would become the first European to cross the continent to the Pacific, beating Lewis and Clark by more than a decade. (The feat had a little-known anticlimax: in order to reach Pacific tidewater Mackenzie had to abandon his canoe and hike for two weeks, and then borrow a canoe from local Natives for the home stretch.)

His epic 1789 voyage down the Slave and Mackenzie Rivers to the Arctic Ocean demonstrated that no feasible northwest passage existed, and his great Pacific voyage showed there was no practical route to and from the Pacific. Both achievements were hard lessons, gleaned by sheer force of will across a landscape that remains mostly off the grid and outside of popular imagination. But despite his great accomplishments, Mackenzie has not gotten his full due. According to historian and *Peace River Chronicles* editor Gordon E. Bowes, Mackenzie's prescience in recognizing the commercial opportunity represented by China in particular—his idea to establish a North Pacific commercial empire to trade furs and fish with Asia, which was ignored by his contemporaries—was centuries ahead of its time.

Distracted by our "Mackenzie moment" feeding frenzy, Ben and I neglect to scout out a camp before dark. A raised clearing on the north bank of the river close to the Halfway River Bridge looks promising, but Ben balks at the thought. It is a place of bad omens: when he visited this very spot a few months earlier, he found a decapitated wolf left to rot; not far from that was the bullet-riddled corpse of a deer, which had

A lone coyote crosses farmland near the edge of the Peace River on Bear Flats during the winter of 2015.

fallen from the bridge above. Like the ungulates that seek out a buffer zone against the predatory unknown, we feel safer sleeping on the islands.

It is pitch-dark by now. There comes a crescendo of rushing water, sucking our canoe closer to the roar as if it is caught in a tractor beam. We are accelerating at an alarming rate, so I stop paddling and hold an LED flashlight like a headlight, letting Ben steer. There is a loud crunch as our canoe scrapes against the rocky bottom, turning us sideways to the current. Our overloaded canoe is in danger of capsizing. I consider jumping overboard and dragging the canoe through the shallow parts, but the riverbed is pockmarked by deep pools. Ben steers randomly for the closest shore, and the canoe doesn't so much land as crash.

It is too dark to discern whether we have landed on the mainland or an island. Outside of our LED light beam, all is black. A weird, unvegetated depression runs parallel to the shoreline, bearing fresh tracks of moose, deer, elk (we think), bears, wolves and/or coyotes. Lynx and fishers are distinct possibilities too. It appears we will be forced to camp on top of the busiest game superhighway in the Peace Valley—on a piece of land apparently touching the densely forested, unpopulated south bank of the river. There are no large trees in the immediate vicinity from which to hang our food, so we leave our feed bag in the canoe.

The kinds of predators that might sniff out our food on this night form a pretty expansive list. Next to black bears and the occasional grizzly, wolves are the dominant predators in these parts, followed by their closely related cousins, the coyotes. Lynx, mink, martens, weasels and otters are known to inhabit these parts as well.

It's no coincidence that this list includes many animals important to the fur trade, which has greatly waned but never quite ended here in the Peace Valley. Today at least nine traplines touch the north and south banks of the Peace River along our route between Hudson's Hope and the Moberly River confluence—just a fraction of the land registered to about 3,500 trappers province-wide. (About half are aboriginal.) Trappers harvest the fur of seventeen species as regulated by the province, which earns a royalty on every retained pelt or skin, from $0.04 to $0.07 for a squirrel or weasel, $7.31 for a wolverine and $7.55 for a bobcat (smaller than a lynx, double the size of a house cat). A black bear pelt is a relative bargain at $3.50.

Vic Gouldie at his home in Hudson's Hope. The sixty-eight-year-old trapper has relied on a lifetime of work building dams to raise his family.

The trapline along the stretch of water we have just paddled—on the north shore of the river roughly between Hudson's Hope and the Halfway River—is registered to Vic Gouldie, age sixty-eight, who has trapped this area since 1974. When we talk later, he makes a point of setting me straight on the popular misconceptions of trapping. A trapline is not just a trail through the bush, he says; it's a large block of territory, in his case about 775 square kilometres. And a serious trapper is by definition a steward of the land. He compares real trapping to good farming, where you work a certain section for a while, then leave it fallow.

Right: Marten that have recently been skinned for fur on the West Moberly First Nations reserve are left on the side of the road for scavenging animals and birds.

Below: Vic Gouldie shows off furs he has taken on his trapline, which touches the banks of the Peace River. He stands to lose a chunk of his trapping territory due to the building of the Site C dam.

"The reason these traplines were registered in the first place is so they won't be over-trapped by every idiot that comes along, as soon as the price of fur hits the ceiling. You kind of manage the place, eh?"

Gouldie won't name his most lucrative species, but others say lynx and martens are important to the bottom line. That said, abundance and market prices are always in flux. Take the lynx that occupy this valley: there is a direct link at any moment between the health of the snowshoe hare population (the lynx's preferred food) and the number of lynx available to be sustainably trapped. There is a roughly ten-year cycle where hare populations boom and bust, which is followed in lockstep by the lynx.

Gouldie is an intriguing character, beyond his talent for fitting an expletive into almost every sentence he forms. He will lose part of his trapline to Site C, but he is not against the dam. In this way he represents a lot of working people in the Peace and far beyond, who ultimately see Site C as a good thing or, at minimum, not the disaster so many of the locals consider it to be.

Before he was a trapper, Gouldie drilled tunnels for hydro dams across the province. He worked on Site Zed on the Stikine River, part of a failed BC Hydro plan that included damming the Grand Canyon of the Stikine. He worked on Alcan's Kemano dam, as well as a dam to power the Port Mellon pulp mill near Gibsons. Mica Creek too. On the Peace, he helped build the tunnels for the W.A.C. Bennett Dam and even did exploratory drilling for Site C the last time it was being proposed.

As someone who has relied on a lifetime of dam construction work to raise a family, he reserves a lot of rancour for the urban environmentalists who descend in droves on the Peace River each summer to celebrate the David Suzuki–associated Paddle For The Peace. "These fucking idiots who come up here to save the Peace, they have this big rally and weenie roast"—here he starts mock crying—"'Oh, we shouldn't, it's gonna be gone forever!' It will be a big improvement." As Gouldie sees it, by the time Site C is complete, which is projected to be 2024, there will be three big lakes in this area, just around the time the Okanagan will be getting too crowded. "This will be the Okanagan of the north. People say it will be all fucked up, and it will be. Maybe for my lifetime, but not for yours. If you have property up here, you'll be sitting on a gold mine." His single concession to environmentalists is that their pressure is forcing BC Hydro to do a better job of building Site C than it did on dam construction in the past, like making sure enough timber is removed from the land in advance of the flood to prevent the future reservoir from becoming a wasteland of woody debris.

He shares his vision of the post–Site C future. Gouldie wants to cut a trail to a beautiful flat bench on his trapline, a spot he has already scoped out, just above the reservoir's high-water mark. There he could build a main cabin, smaller outbuildings, a dock for tourist fishermen, and even horse trails for the kids. "Now that's positive thinking!" he growls.

The only furred animals we end up seeing on this night are beavers. Lots of them. They are so numerous on the shoreline that every piece of firewood we collect is already pre-trimmed to size, the conical tips covered with gnaw marks. We would go on to see at least fifty dams during our four-day river passage, and without even looking for them. The smacking sound of the beavers' flat tails on the water—seemingly to warn the others of our presence—is unnervingly loud, especially at night. Trapper Vic has a different explanation for the tail smacks: they are not alerting other beavers to my presence, but testing my reaction to a very loud, sudden noise. A

We saw at least fifty beaver dams like this one during our four days on the river. The smacking sound of the beavers' flat tails on the water was unnervingly loud, especially at night.

predator wanting to eat the beaver would react very differently than an animal minding its own business.

Among the largest rodent species on the planet, beavers were what drove men like Alexander Mackenzie to the Peace Country in the first place, not in the name of nation or empire, but for a loose association of European and Canadian businessmen called the North West Company, then engaged in a cutthroat battle with its hated rival, the Hudson's Bay Company. Hats made of beaver felt were a status symbol in Europe for about three hundred years, beginning in the late sixteenth century. The beaver's recovery from being widely trapped to extirpation to make such headwear—including the navy cocked hat style famously worn by Napoleon—speaks to an uncanny resilience.

Beavers drove fur traders into the Peace Country in the 1790s.

Thinkstock/iStock/MyImages_Micha

The building of multiple hydro dams on the Peace, including the daily rise and fall of the machine river, has been less of a problem for the beaver than I presumed. Michael Church, a University of British Columbia river expert and professor emeritus who has studied the Peace for decades, says beavers have benefited greatly from the elimination of the annual spring freshet. There was a big explosion of beavers after the construction of the W.A.C. Bennett Dam in the late sixties, and Church should know. For many years he has been cursing the "damn beasts" for felling trees across the riverside areas he has been trying to study.

After a dinner of rotini and Bombay Sapphire, we retire early on this night. Hemmed in by towering grey cliffs, this spot is a natural amphitheatre for every possible animal sound, real and imagined. Hours later, out of absolute silence, a coven of wolves starts howling. We're later told it was likely coyotes. Either way, it is a bloody canine opera: most of the high-pitched yaps are dog-like, with the exception of a single werewolf-like howl that rises quickly and lingers on the fall. The sounds bounce across the valley, simultaneously near and distant; it is enthralling and terrifying at the same time. The party eventually dies down, as the white noise of the rushing water that stranded us here drowns out everything else.

DAY 3 | LOST RIVERS, TIMBER CRUISERS AND A BLACK BEAR

WE WAKE UP on a tiny sliver of an island: it is sunny, unseasonably warm and much too quiet. I walk out onto the shore to discover our canoe sitting in a channel of rocks. The water has disappeared. It's our first direct introduction to the Peace as a remote-controlled machine, where water levels can rise or fall over the course of a day, creeping up on the island's shoreline to a barely visible permanent grassline, only to recede hours later, exposing sandy river bottom.

Right: I drag the canoe through low water on the Peace River after our second night of camping. The fluctuation of the water was caused by a drawdown from the existing dams on the Peace River.

Traffic rises and falls on Highway 29 just north of the Peace River at Bear Flats.

The amount of water flowing on the Peace at any given moment is a direct reflection of electrical demand farther south, toward Vancouver, where more than 50 percent of the province's population lives. At times of higher demand—such as during evenings and when winter heating is most required—more water is released through the dams to generate electricity; periods of lower demand, like weekends, see more water stored in the reservoirs. In this way, the reservoir behind a dam is like a giant battery, storing water to run through the turbines when it is needed.

After breakfast I drag the canoe half a kilometre to the tip of the island, where the dewatered channel trickles into the main stem of the Peace. The water is so cold it hurts, but I don't mind the toil: for months I have been chained to a desk reading about the exploits of voyageurs, so it's about time I do some portaging of my own.

Bear Flats as seen from the Peace River.

Our rented canoe scrapes along the rocky bottom for the next twenty minutes, leaving a trail of red paint on the rocks, before she is liberated into deeper water.

Today we hope to cover about 25 kilometres of the river. If we can make this distance, we will be close to the first construction site for the new dam—just above the point where the Moberly River meets the Peace, upstream of our haul-out at Taylor, BC. Within the next day, our goal is to land on the recently clear-cut island located where the actual dam will stand.

As we approach the high bluffs of Bear Flats, the highway veers away to the north, leaving us truly alone for the first time.[7] Site C feels a world away from this point. The only sounds now are birds and the occasional lowing of beef cattle foraging on Crown land. To the north of the river, a yellow-green mosaic of grassland, farms and forest

An anti-dam sign visible from Highway 29, on the edge of fields owned by Ken and Arlene Boon.

The front gate at the home of Hudson's Hope mayor Gwen Johansson. As the signs indicate, her property will be completely submerged by the Site C reservoir.

Photograph courtesy of Emma Gilchrist, DeSmog Canada

reaches into the heart of BC's oil and gas country. The north and south banks of the river get completely different sun exposure, with each hosting a different array of ecosystems. It's on the south-facing slopes that the prickly pear cactus grows, an aberration for a river valley this far north. With more solar heat, rare grassland ecosystems thrive, attracting elk and mule deer. The valley microclimate also attracts southern species like robins, which must puzzle at what drew them this far north during the late spring snowstorms known to strike in these parts. Meanwhile the north-facing slopes of the river are more typical of a valley in the boreal zone, with vast stands of conifers dotted with deciduous trees like trembling aspen.[8]

On the manicured plateau we see the barns and prosperous farmland owned by Ken and Arlene Boon, third-generation landowners who we will meet after landing at Taylor. Barely visible through my binoculars are the signs they have put up on their land, marking the level that the Site C flood waters will reach. (They stand to lose everything.) One sign combines hard advocacy with entrepreneurship, all in giant bold caps: "SITE C SUCKS. HAY FOR SALE."

Our first destination early this day is Rocky Mountain Fort, now an archeological site—the first European settlement on the mainland of British Columbia, established the year after Mackenzie blew through on his way to the Pacific. The site was repeatedly occupied, moved and abandoned over the decades.[9] My interest in the fort is that it served as the forward outpost for fur trader/explorer John Finlay, whose name was given to the Peace's mightiest tributary, which was indelibly altered by the creation of the Williston Reservoir. Using Rocky Mountain Fort as a forward base, Finlay set out in 1797 to explore this great Peace tributary—one of the wildest, most treacherous in the province, whose headwater lake is also the ultimate source of Canada's largest river by drainage area (just smaller than that of the Yangtze of China), the great Mackenzie.

A young Frank Swannell in action in 1910, in north-central BC's Endako River valley.

Image I-33576 courtesy of the Royal BC Museum and Archives

The first complete, organized exploration of the Finlay is credited to Samuel Black, who in 1824 survived its great canyons, steep cataracts and impassable hazards, confirming without a doubt that the river represented a barrier to navigation. He was followed by Frank Swannell, British Columbia's great surveyor and cartographer, who covered a large portion of northern BC in his work for the provincial government from 1908 to 1939. It was Swannell, among the last of the New World explorers, who led a harrowing 1,900-kilometre exploratory ascent of the Finlay by canoe in 1914, eventually surveying and mapping the entire Finlay River and the Peace as far east as around Hudson's Hope.

The goal of the Finlay expedition—with a total of six men in a huge cottonwood

A crew runs the rapids of Deserter's Canyon on the Finlay River in October 1914, a feat photographed by Swannell himself.
Image I-33205 courtesy of the Royal BC Museum and Archives

log they had chipped and charred into a dugout canoe—was to survey all the way to the headwaters in the Stikine country. Recounting this experience in the *Beaver* in 1956, Swannell recalled how his team got stuck at a spot called Deserters Canyon for days, forced to circumvent a fearsome whirlpool that could inhale huge spruce trees and spit them out hundreds of metres downstream. Along the way they found evidence of the unfortunate Klondikers who had been foolhardy enough to run this river less than two decades before: evidence of the great forest fires they set, and the skeletons of their pack horses, saddles still in place as they died of exhaustion or were shot. Running low on supplies in the upper Finlay, the party was forced to live off the berries, fish, rabbits, beavers and grouse they could kill. Multiple harrowing canyon passages later, James Alexander, a strapping mixture of Scottish and aboriginal stock from Fort St. John, shot a moose as it swam in the distance, but he only wounded it. Desperate for a steak, he dove into the water and sliced the moose's throat, riding the dying animal to shore as it bled out. The survey ground to a halt just 50 kilometres away from Thutade Lake, the headwaters of the Mackenzie; the team faced a critical shortage of food and deterioration of both the weather and river conditions. Fifteen years would pass—including four years of fighting in France during World War I—before Swannell would return and at long last reach and map the headwaters.

Frank Swannell is unsung in British Columbia, in the sad way that Canadians fail to acknowledge their homegrown heroes. He would think nothing of leaving in advance of his exploring party to climb a nearby mountain alone for a week, sketching and taking measurements of the surrounding country, armed only with his instruments, a bit of fishing line, and a pack of dry rice and tea. It took essayist Edward Hoagland, an American, to track down one of the world's last great living explorers (in a Victoria seniors' apartment in 1966), to talk to the man who surveyed and mapped more of central and northern British Columbia than anyone else. The meeting, which was strained and uncomfortable, appears in Hoagland's classic *Notes From the Century Before*. Swannell, who was eighty-six by this point, found his brash American interviewer's stutter disturbing. Although Hoagland was star-struck to be "sitting next to the equivalent of de Soto," he failed to get Swannell to open up and provide the kinds of human stories he sought for his book. "He was a technician," concluded Hoagland, "a narrow man, whose memories were never of people, only of terrain."

Hoagland had by this point come to see British Columbia as a "paper province" because much of it, largely uninhabitable wilds like the region flooded by the W.A.C. Bennett Dam, was rarely if ever traversed by any human being. It existed only in the abstract, he said, in the form of paper topography depicted by maps made by rare men like Swannell—one of the few Canadians daring enough to tread there.

The years leading up to Swannell's epic Finlay expedition saw great changes to the Peace region after many years of limited development. On most of the land from the Rockies east to the Alberta border, homesteading had been prohibited right up to

Fertile Valley of the Peace and Halfway River

AND THE LAST FRONTIER OF THE

GREAT WEST

The Most Talked About New Country of the Day — The Garden of the West

THE soil is rich and deep, all farm products of the temperate zone are grown, as well as the finest grain in the world. There are indications of Coal, Oil and valuable Ores in different sections of the district. The country is waiting—waiting for transportation. This it will have shortly, and within a few years the most inaccessible points will be within easy reach. THE TIME FOR PROFITABLE INVESTMENT IS THE PRESENT.

See— **S. P. DUNLEVY**

301 Dominion Trust Building Phone 5753 Vancouver, B.C.

Reproduced from *British Columbia Magazine*, June 1911

A developer land ad, reproduced from *British Columbia Magazine*, June 1911.

Image from Gordon E. Bowes, Peace River Chronicles *(Vancouver: Prescott, 1963), 301.*

the eve of World War I. (It was held in a reserve established by the provincial government in return for aid given by Ottawa for railway construction elsewhere in the province.) When the federal government finally opened up what was known as the "Peace River Block," farmers poured in, establishing farms around places like Dawson Creek. They quickly discovered it to be good land for grain crops, hay and, in the remarkable valley, where its fertile alluvial soils are nourished by warm Pacific air and the annual freshet, vegetables and tree fruit.

Prior to World War I, at least twenty-four charters were granted for railways that would connect the Peace to the outside world; it would take until 1958 before such a link was built. This remoteness and lack of highway and rail connections meant the Peace remained the domain of the trapper and frontier farmer well into the twentieth century. It was, in the words of one developer, the "last frontier of the Great West."

Our search for Finlay's forward base, Rocky Mountain Fort, leads us a few kilometres north of Bear Flats, where Tea Creek, a trickle of orange pekoe–coloured water, runs into the Peace. We land and begin exploring the flat shoreline for the sorts of disturbances that would denote the presence of the settlement. We find nothing. Tracing the creek inland around a sharp bend hemmed in by cliffs of red clay, we run into a black bear.

My mind jumps to a news story I read before leaving Vancouver. In May 2015 a twenty-seven-year-old man named Daniel O'Connor, a resident of nearby Mackenzie, was sitting alone before a campfire on a forest service road near the Parsnip River. He was last seen at about 1:30 a.m. At some point in the night, a black bear attacked him and dragged him 200 metres into the forest. His body had no defensive wounds; the bear, which the RCMP later shot, had stalked and killed O'Connor for food. Such an attack is among my greatest nightmares: a hungry bear intent on making me his dinner.

Opposite: I cautiously walk up Tea Creek after spotting bear tracks.

The black bear ahead is about ten metres away, and not very threatening—in fact it appears to be dozing by the streamside. Ben and I freeze. "If it wakes up or notices us, whatever you do, don't run," I whisper. "They're hard-wired to chase anything that flees." Ben looks at me quizzically and I read his mind: he's six-four, all legs, and could easily win a sprint to the canoe.

Experts say bears will attack when they are surprised, to protect their young and sometimes for no discernible reason at all. And if it intends to eat you, a bear will show telltale signs: it will approach with its head lowered, its ears pinned back, making intense, constant eye contact with the intended quarry. Even when you do everything right—including making verbal noise, avoiding eye contact and making your body as large as possible—it will not stop advancing. Bear expert and author Brad Garfield says different bear species will stalk and attack a victim differently: a polar bear is slow and patient, while grizzlies are all speed and unstoppable brawn. A mature male grizzly can stand up to three metres tall on his hind legs and weigh 700 kilograms; despite its size, it can close the gap at over 55 kilometres per hour. A black bear will typically be more sneaky, moving in zigzag patterns if space allows, hiding behind trees and obstacles in hopes of attacking by surprise. Or in our case, it will just lie there and ignore you.

A bear print on the banks of Tea Creek. New road access from Site C construction will be an issue for black bears in this area, resulting in increased hunting and illegal poaching.

Black bears (*Ursus americanus*) like this one are plentiful in these parts. They are master opportunists—omnivores that are not picky about what they eat, and therein lies their success. They have adapted to Louisiana swamps, coastal forests of British Columbia and Alaska, Mexican scrub forests and Labrador tundra. In spite of extirpation in some parts of the US and Mexico, black bear populations are ascendant across much of the continent. According to the International Union for Conservation of Nature, there are nearly a million black bears in North America.

BC Hydro has noted that increased road access from Site C construction will be an issue for black bears, resulting in increased hunting and illegal poaching. Meanwhile, the fifty islands that will be destroyed are critical calving habitat for all the large ungulates that bears love to eat—moose, mule deer, elk and more. Anything that causes

Following Tea Creek northward away from the river, we turn a corner and encounter this black bear. We initially assume it is taking a nap.

a decline in big ungulates will eventually affect black bears, and the grizzlies that visit the area as well.

A bear will sometimes doze after a kill the way I nap after Thanksgiving dinner, and we suppose this is the case here. We look closer through one of Ben's lenses, and the bear seems strangely sunken into the clay streambed, as if dead. I hold up a rock and throw it. The bear doesn't move. We approach tentatively. I clutch a driftwood club as Ben wields a canister of pepper spray (the label says it's good for a maximum of three seconds of steady spray!). The bear is dead for sure, but we do not see any signs of trauma on the body. Its fearsome canine teeth are bared, as if it's snarling at the ants in the mud. It's also a relatively young bear, a fact that raises another alarm: the possibility of a distraught mama lurking nearby.

Something large is crashing through the underbrush around the corner. It's getting louder, and time drags as we stand our ground. Two powerfully built men appear, the sun flashing off their neon yellow and orange safety vests. The two youth, not a day over twenty, are "timber cruisers"—members of an invading gumboot army of recent tech school grads who have descended on the Peace Valley in recent years, working as consultants for BC Hydro and a galaxy of established and ad hoc consultancies conducting all manner of studies spawned by the Site C dam. One of them is friendly. He says they are employed by an environmental consulting company based in Smithers, on behalf of unnamed forestry interests. We are told companies are poised to log the entire valley. It's their job to survey an area the size of twenty Stanley Parks (which is about 80 square kilometres in size), highlighting the best and most accessible timber. A big part of what they are doing is determining which areas can be accessed by heavy machinery, such as feller bunchers, which have revolutionized forestry in parts of the province. (A single feller buncher can replace the work of a dozen seasoned, chain-saw-wielding lumberjacks.)

There are at least 1.2 million cubic metres of merchantable balsam poplar, white spruce, trembling aspen, lodgepole pine and larch in this valley. The white spruce is the big prize, says the surveyor as he points to the thick conifers that surround us. All the trembling aspen, especially big in this riparian zone, is destined for plants in Fort St. John and Dawson Creek that make oriented strand board (an engineered particleboard); sawmills like the huge one in Fort St. John stand to benefit from the influx of conifers. The other surveyor, more

The north side of the Peace River about ten kilometres east of Hudson's Hope.

aloof than his companion, cannot understand why we are here. "I'm a freelance journalist," I say, a bit irritated. "And Ben is a photojournalist. We are here to document what's here before it's gone."

And so we stand there awkwardly for a moment or two, staring down at the dead bear. The consultants look on as we roll the animal over, looking for trauma or bullet wounds. Nothing. (Rigor mortis has set in; it looks and feels as if a taxidermist has already preserved it.) It is not malnourished either, but a healthy size. We later learn that anywhere from a quarter to half of all black bear cubs will die within their first year of life, mostly from natural causes.

"It's crazy," says the friendly one, motioning around the densely forested little valley carved by the tributary stream. "You feel bad that this is going to be underneath 50 metres of water." Then he brightens, as if an idea suddenly came to him. "There's years of work here though... if it goes through."

On several occasions he uses the same phrase—"*if* Site C goes through." That's because even though BC Hydro has commenced construction, a degree of uncertainty surrounds the future of the project. Of multiple lawsuits active in the courts at this point—brought by various First Nations and landowners—most have failed by summer 2016, but at least three appeals remained undecided.

BC Hydro's strategy has been to surge forward with construction and contracts, creating an air of certainty around the dam regardless of potential roadblocks. In December 2015 the utility announced it had reached an agreement on a $1.75 billion contract with a consortium to build the earth fill dam, spillway, diversion tunnels and powerhouse foundations. About four months later, a $470 million contract was awarded to design, supply and install turbines and generators.

A huge point of contention has been what all of this will end up costing. By the time BC Hydro filed its environmental impact statement in 2013, the cost had risen to $7.9 billion. When the project was finally approved a little more than a year later in December 2014, the price tag was $8.335 billion, with a $440 million contingency, to total just under $8.8 billion.

Viewed globally, overruns on cost estimates for mega dams are routine, both in the past and present. A 2000 study by the World Commission on Dams surveyed over eighty large dam projects, finding an average cost overrun of 56 percent. In 2013 an Oxford University research team followed up on that study, looking at 245 big modern hydro dams in sixty-five countries. They concluded that dam cost estimates are as wrong today as at any time during the seventy years for which data exist: the construction costs of large dams are on average about 90 percent higher than their budgets at the time of approval, in real terms.[10] Long construction windows also make the projects vulnerable to factors such as hyperinflation, swings in water availability and electricity prices. Study co-author Bent Flyvbjerg suggests that as a general rule, smaller and more flexible energy projects can be built and brought online quicker, while also being more easily adapted to social and environmental concerns, than "high-risk dinosaur projects like conventional mega-dams."

We walk with the surveyors back to shore and their waiting jet boat. We never find any trace of John Finlay's forward base because it was never here. The remains of the outpost in question are in fact just upstream of where the Peace River meets the Moberly River. Located at almost the exact point where Site C will cross the river, the archeological remains of the fort were destroyed in early 2016 to make way for the dam.

Back on the water, we choose a camp for the night: it's a beautiful spot, high up, overlooking the north bank of the river, with a huge picnic table and lots of wood and dry kindling. (It's maintained by the River Rats, a local volunteer group.)

Ben's fly-fishing technique off the island bank is primitive—there's none of that gentle placement of the fly on the water, none of the delicate artistry—but he's starting to catch more fish than I am. First a nice whitefish, and then in quick succession a bull trout and another species of whitefish. It's tempting to keep these for eating, but before I can suggest this to Ben, he has released his second whitefish back into the river. It's the last fish either of us will catch on the Peace.

DAY 4 | SITE C BECOMES REAL

TODAY IS THE HOME stretch: we are less than a day's paddle from our final destination of Taylor, just south of the natural gas hub of Fort St. John. Before we get there, we will catch our first glimpse of the Site C construction, which began the day we launched from Hudson's Hope.

We amuse ourselves for a spell on the coast of our private island, watching juvenile bull trout hide in slack water along the shoreline. They repeatedly ambush and inhale mayflies as the current pulls the bugs past their hiding spot. Twenty minutes later they're still at it, like gluttons at an all-you-can-eat buffet who never get too bloated to quit.

Right: BC Hydro's Site C dam site directly across from the Moberly River confluence. Construction started the week we paddled the Peace.

CAT

We pack up and get back on the water. By noon a dull roar is audible upstream, accented by the rising whine of hydraulics. The sound is incongruous with the river in its stunning autumn colours. "It's the rumble of progress," mumbles Ben.

The din grows as we pass the confluence with the Moberly River, which we never actually see—it is hidden from view because we have paddled between the north riverbank and an island where the clearing work has begun. The fate of the unseen Moberly, a major tributary, is a big issue with local First Nations here, largely because the waterway supports a diverse community of fish, including Arctic grayling, mountain whitefish, northern pike and a weird thing called burbot. Known as eelpout, ling and "the lawyer," burbot are prehistoric throwbacks, resembling something between an eel and a catfish. These mysterious bottom-dwellers hunt at night, stalking prey by smell and by vibration (the latter detected by the burbot's distinctive soul patch "barbell," located on its lower lip). They are known to swallow whole whitefish, kokanee and suckers almost as large as themselves. While burbot are not considered at risk in British Columbia, there are concerns that hydro dam construction has affected their numbers on the Kootenay River and on the Columbia downstream from the Hugh Keenleyside Dam. Maybe one day here, as well.

Opposite: I wander a clear-cut island located at the Moberly–Peace River confluence. It's at this very point that the Site C dam is planned to stand.

On the north shore, seven big pickup trucks are parked, with two jet boats lashed to the bank. We paddle close to the shoreline to catch the express currents, just as a big front-end loader breaks through the edge of the forested bank. From my

vantage point, a thick cottonwood is about to fall across our canoe. "Ben look out!" I make eye contact with the operator who nearly jumps out of his seat in surprise.

We make for an adjacent island, where hundreds of alders and thick cottonwoods have been clearcut. This is the exact point where the Site C dam will throttle the Peace River main stem—the hard edge of a future reservoir, which will presumably bear the name of a BC politician: Clark Lake and Recreational Area? Bill Bennett (The Lesser) Reservoir? As Ben snaps shots of the construction, a pack of BC Hydro contractors stand about ten metres away, grim-faced and staring.

The plan to build Site C will unfold something like this: it will take nearly a decade, with most of that time involving excavation and material relocation. Coffer dams and diversion tunnels will divert the main stem flow of the Peace for several years to allow completion of the generating station and dam structure. BC Hydro consultants have estimated it will take about three months to drown the 9,300-hectare reservoir area, followed by fifteen months of fine tuning before the dam can go into production. If all goes as planned, it will be running in 2024.

Work crews at BC Hydro's Site C dam site on the north bank of the Peace River.

Site C will be a very different project than the W.A.C. Bennett Dam. For one thing, it will technically be a run-of-river hydro project, because although it will have its own reservoir, it will rely primarily on the four-armed aquatic monster of Williston Reservoir for storage. This in particular is what makes Site C a green project to BC Hydro: they claim Site C will be able to generate 35 percent of the energy of the W.A.C. Bennett Dam, with just 5 percent of the reservoir footprint. But the sight of the first clear-cut demonstrates how subjective the word *green* can be. Not only will the creation of a 9,310-hectare reservoir generate methyl mercury that will persist and rise up the food chain, the deluge will also see a spike in greenhouse gas emissions that will not return to current conditions for decades. The decomposition of soil and vegetation in the reservoir will produce methane, which has an estimated global warming potential up to thirty-nine times higher than CO_2 (which the project will also produce).[11]

So being green is all about trade-offs. In building Site C, the government is concentrating its energy-related impacts on a single river. Instead of building a mass of smaller hydro projects with cumulative impacts on salmon rivers across British Columbia, for example, they are concentrating their efforts on a river already compromised—which sounds like a good thing, until you consider how vulnerable this will make the future energy supply should climate change substantially affect the amount of water that spills into the Williston Reservoir. Considering that the existing Peace River dams already supply the province with about a third of its total electrical capacity (and more for energy), is it prudent to grow our dependence on a single river system?

The final word on whether or not Site C is actually green goes to California. This state has historically been one of BC's prized energy export markets, and the last time Site C was proposed in the mid- to late 1980s, California was the export target. But today, hydro projects bigger than 30 megawatts (MW) do not even qualify as renewable energy under California's current renewable portfolio standard (RPS) requirement. And while solar and wind generation surge, even small hydro is expected to account for less than 0.5 percent of the state's renewable, online capacity by the end of 2016. To California, Site C is a dinosaur—and anything but green.

Back in the canoe, we flow along the north mainland, with Ben snapping as I keep the canoe even with the bank. A large white pickup has been following us for at least ten minutes now, driving a new road cut along the river. Two men in white hardhats jump out and walk to the shore just metres away from us as if they want to talk. We wave. They stare back.

The rumble is behind us now. We round the bend in the river to encounter two men in their underwear, standing behind a jet boat. The latest drawdown of the river has trapped them on a rocky shoal of ankle-deep water. Both are in the river pushing: one is young and strong, the other is in his seventies. (The older man is deathly pale and looks like he's on the verge of a heart attack.) We spend the next thirty minutes heaving the jet boat over wetted rocks toward deeper water. Eventually they have enough clearance to float.

Two men are trapped in low water just downstream of the Site C construction site. The fluctuating river levels that result from generating electricity mean that boats can get stuck on the rocks.

C-GPBA

We paddle on. This is our victory stretch, the culmination of four days on the river—no small achievement for two soft guys from the city. But no. About a kilometre above Taylor, a helicopter circles our canoe and slowly lowers over us. Two grim-faced men stare downward; one shuffles papers and points to each of us in turn. The helicopter levitates so closely that the river surface becomes boiling whitewater. Then as quickly as it appeared, the helicopter disappears over the southern bank of the river. Two men emerge from a clearing on the same coastline about five minutes later, waving and shouting for us to stop. They are a helicopter crew from North Peace Search and Rescue (NPSR)—the advance unit of a large-scale manhunt that has been under way for hours. The search started at Hudson's Hope, the last place we were seen alive.

"Are you Christopher Pollon and Ben Nelms?"

At this moment a fixed-wing plane and pack of good Samaritan farmers in jet boats are also searching the Peace River, looking for two big city journalists in a red rental canoe. Slowly we figure out what has happened. Ben neglected to keep his wife posted on our paddling plan and underestimated how alarmed she might become if we went a few days without checking in. Three days without word and she called in the heavies.

Opposite: North Peace Search and Rescue descends upon two oblivious paddlers.

The manhunt is officially called off. Emergency Management BC statistics later show that Ben and I constituted the only "inland water" search and rescue incident in NPSR's entire 33,000-square-kilometre region for the week of September 20–27, 2015. One of the men turns as he walks back to his helicopter: "The next time we get a call about you guys, you're paying."

Our haul-out point on the river—Peace Island Park at Taylor—is immediately downwind of the McMahon Gas Plant—a miniature *Lord of the Rings* Mordor, complete with a tall tower with a giant flame burning at the top (minus the all-seeing eye wreathed in fire).

In a strange way, this sprawling plant—run by Spectra Energy of Houston, a Fortune 500 company that processes and transports

to market much of BC's natural gas—represents an alternative to Site C. More than a processing facility, the McMahon plant is also a 120 MW co-generation plant, powered by local natural gas. Commissioned in 1993, it supplies electricity to the BC grid, and at the same time provides steam to power the facility's operations.

The debate over possible alternatives to Site C—and by extension whether the dam is even required—has grown in volume since the project was given the green light in December 2014. Even the Joint Review Panel entrusted by BC and Canada to investigate the environmental, economic, social, health and heritage effects of it concluded that BC Hydro "has not fully demonstrated the need for the project on the timetable set forth."[12] In March 2015, about three months after the province decided to move forward, panel chair Harry Swain, a former deputy minister of Industry Canada and Indian and Northern Affairs Canada, took the unprecedented step of speaking out as a private citizen. "There's a whole bunch of unanswered questions, some of which would be markedly advanced by waiting three or four years," he told *DeSmog Canada*, a Victoria-based online news magazine that has led provincial media in scrutinizing Site C.

Opposite: The north side of the Peace River approximately ten kilometres east of Hudson's Hope, just before we paddled through the Gates.

If Site C currently lacks a social licence in British Columbia, it's because critical questions about alternatives and project need have not undergone a fully transparent, independent review. In its Clean Energy Act of 2010, the BC Liberals exempted Site C from requiring a Certificate of Public Convenience and Necessity—which is otherwise mandatory and only granted after a lengthy, in-depth project review by the BC Utilities Commission (BCUC), which acts as an independent regulator of BC Hydro. This quasi-judicial body, self-funded by levies against the utility, has a mission to ensure BC hydro shareholders are afforded a "reasonable opportunity to earn a fair return on their invested capital."

When Site C was proposed in the early 1980s, the Utilities Commission was allowed to do its job: reviewing in minute detail the project design, impacts and ultimate justification. Performing due diligence on such a huge public expenditure was a major undertaking: a twenty-five-month process by a five-person panel, with 116 days of formal hearings. Separate phases scrutinized BC Hydro's proposed project cost and forecasted future demand. Regarding the latter, the utility at the time based its case for Site C on its 1981–82 forecasting, which predicted that BC Hydro's system-wide energy demand would grow to about 60,000 GWh/year by 1992–93. By the time the process was wrapped up in May 1983, the commission concluded this forecast was optimistic and recommended the provincial cabinet defer issuing Site C a certificate. The government accepted the recommendation.

Time has proven this independent scrutiny served a critical fail-safe role in protecting BC Hydro customers from a project they did not need. Consider that the nearly 60,000 GWh/year system-wide energy demand forecasted by BC Hydro in 1981, which they said would be needed by 1992–93 as its justification for building Site C, is only approaching reality in 2016–17. But this time around, Site C has been exempted from Section 45 of the Utilities Commission Act, which would have necessitated an independent review. The government ignored Swain and the joint panel's final recommendation for a BCUC review as well.

The Joint Review Panel came at the tail end of a multi-year federal-provincial environmental assessment (EA), and was forced to operate in a streamlined nine-month time frame that was incapable of making key determinations about project cost and need. During this fast-tracked process the panel was expected to coherently review at least 24,500 pages of information. Swain later told me they had no means of examining or challenging the capital cost estimates of Site C or its alternatives. "They just asserted that KPMG had looked at it and it was good, and they never did provide the detailed calculations. There was no discussion of the ensuing rate consequences, or debt for the provincial government, or financial position of BC Hydro. These are all things that the Utilities Commission would routinely examine."

To determine whether the $8.33 billion cost estimate was sound, the Site C budget was instead reviewed by a team of construction and engineering experts and internally within government, with an external review by tax/auditing company KPMG, which was refreshed again in 2014. When pressed on why the government exempted Site C from Utilities Commission scrutiny, Minister of Energy and Mines Bill Bennett said the following in September 2015: "There was a decision made that if government was to build Site C, it would be a monumental decision in terms of energy policy that only duly elected officials have a right to make, as opposed to [an] organization like the BCUC that is made up of bureaucrats and lawyers."

During the joint panel review, BC Hydro revealed that Site C was projected to lose about $800 million in its first four years of operation beginning in 2024—meaning the electricity would be surplus and sold at a loss. (Swain confirmed this with his own calculations.)

When we talked in summer 2016, Swain dropped a bomb, estimating Site C power will not be needed until at least the 2040s. That's because BC Hydro's 2012 demand forecast, used by the utility as a justification for the project, asserted a demand growth of 2 percent per year, which has not materialized. In fact, demand for power in BC has already been flat for over a decade, due to slow growth in residential and commercial load and a failure of industrial demand. At the same time, the beginning of very steep rate increases is just beginning to be felt by BC Hydro customers. By the time the dam is built, and the Utilities Commission allows the price to be rolled into the rate base, Swain says the real price of electricity will have increased by "something in the order of 50 percent." With electricity prices this high, residential, commercial and industrial BC Hydro customers will reduce consumption and seek cheaper alternatives. "We would expect that such a price increase would have a very serious depressing effect on demand."

In 2013, while the joint panel was continuing to wade through mountains of BC Hydro–generated paper, economist and Simon Fraser University professor Marvin Shaffer, who has focused on energy, transportation and natural resources for the last thirty years, was hired by the Peace Valley Environment Association to examine BC Hydro's case for building the dam. Shaffer's conclusions centred not so much on the Peace Valley, but on an industrial site in the Vancouver suburb of Port Moody. It's there that Burrard Thermal, a 900-MW natural gas–powered thermal station, used to provide an emergency backup for the entire BC electricity system.

As Shaffer explains, in order to have a reliable electricity system, BC Hydro must plan ahead to make sure it can meet demand at all times. This includes planning to ensure that in a worst-case drought scenario (which would sap the hydro dams that provide most of our electricity), the lights could still come on. Burrard Thermal, coupled with our ability to import low-cost power from outside the province, up until recently provided that emergency backup. But Premier Gordon Campbell put BC Hydro in a straitjacket by eliminating both of these sources, resulting in a 900-MW loss to the system from Burrard Thermal alone. As well, the Clean Energy Act of 2010 that exempted Site C from BCUC review also closed the door to buying cheap power on the international spot market, including the possibility of using the already-paid-for 1,300 MW that BC already is entitled to use under our treaty with the US over the shared Columbia River.

BC Hydro needed to replace what was lost with Burrard Thermal, and Campbell turned to the private sector, resulting in the growth of BC's independent power industry.[13] But projects by independent power producers (IPPs) were not a good replacement for Burrard and our previous ability to import from spot markets. They can provide energy, but they lack storage and thus cannot provide a dependable replacement of peak capacity. And unlike in the case of Burrard, which could sit idle until we needed it in an emergency, IPPs have signed contracts with BC Hydro to sell the province power. In average or above-average water years, it's all surplus and we're forced to buy it whether we need it or not.

By 2015, about a quarter of all power generated in BC came from IPPs—that year alone BC Hydro bought over 13,000 GWh of IPP energy at a cost of over a billion dollars. (Shaffer says BC Hydro has had to buy at least 5,000 GWh from IPPs to replace what was lost with Burrard.) It's exorbitantly expensive supply, says energy analyst Arthur Caldicott, who has estimated that BC Hydro will be on the hook for $54 billion in private-sector power contracts over the next fifty-six years.

Instead of Site C, Shaffer has suggested BC Hydro should replace what we lost with Burrard Thermal by building more hydro capacity, by adding new turbines at existing hydro sites and, if required, by building relatively inexpensive single-cycle gas plants for backup and capacity. (The gas plants could be established close to centres of demand and would have low emissions because they would only be run as needed.) That approach could be coupled with importing cheap electricity on the spot market if needed—including wind and freshet-hydro surplus power from around the Pacific Northwest. "With these measures, Site C wouldn't be needed until after 2030 or later, even if BC Hydro forecasts of demand somehow prove correct," he said.

Shaffer is just one of many who has pushed for alternatives to Site C. In May 2016, the Royal Society of Canada (RSC)—a 132-year-old national organization for research and learning—took the unusual step of entering the Site C fray. Joined by 250 legal scholars and scientists, the RSC called on the government to suspend Site C construction until it undergoes further independent review and ongoing First Nations legal challenges are resolved.[14] It was also necessary, the RSC said, to consider less destructive alternatives to the dam. That's because the number and scope of "significant adverse environmental effects" from Site C dwarf any other project ever reviewed in the history of the federal environmental assessment process in Canada. For example, the LNG Canada project proposed near Kitimat was approved in 2015, with greenhouse gas emissions being the sole significant adverse environmental effect. For Site C, twenty-two separate conclusions of significant adverse environmental effects were identified. Beyond this, Treaty 8 First Nations, engaged in unresolved court cases regarding potential infringement of aboriginal and treaty rights, would be affected by impacts to fishing, hunting, trapping and traditional uses of the land.

In a briefing note prepared to support the announcement, the group looked at the three alternative portfolios of energy resources that BC Hydro considered along with Site C (each making up about the same energy and capacity as Site C), and found that one of the portfolios in particular—which included wind energy, turbine additions at existing dams and limited natural gas generation—provided the "most likely alternatives to Site C." It would still potentially meet the requirements of the Clean Energy Act, they reported, including with respect to greenhouse gas emissions and competitive electricity rates, while avoiding an avalanche of environmental impacts.

As we land at Taylor on this rainy night in late September, we are more than oblivious to the political maelstrom brewing over Site C. We have much more immediate concerns: not only has our achievement on the river been utterly deflated by the manhunt, we now face the realization that our truck is more than 100 kilometres to the west in Hudson's Hope.

In that instant, Ben hatches a dubious plan: he will hitchhike in the dark from the nearby highway bridge back to Hudson's Hope for our truck. If he's lucky, he'll be back in about four or five hours. I am to remain on the muddy landing, guarding our canoe and belongings. I think it's a dumb plan but I don't have any better ideas. Ben disappears into the night.

Ken and Arlene Boon's farm near Bear Flats. As of summer 2016, BC Hydro is planning to realign Highway 29 directly through the Boons' house.

PART TWO:
LAND

DAY 5 | GAS AND FARMS, BOOM AND BUST

WE WAKE UP in a Fort St. John hotel to news that we are famous across the Peace Valley. An informal network of local landowners in the valley, many of whom volunteered for the manhunt, are spreading the word that a couple of city slickers "got lost" on the Peace River.

Ben's attempt to hitchhike to Hudson's Hope in the dark was another fiasco. Not only did no one pick him up, but the truckers introduced him to something called *rollin' coal*—slowing as if to stop, then gunning their engines to choke him with noxious clouds of black exhaust. He returned to the Taylor boat launch about two hours later to find me standing in the mud where he left me.

Farmers and ranchers gather at an auction just outside of Fort St. John in September 2015. Although it is known province-wide as a supply centre for oil and gas, Fort St. John has its deepest roots in farming.

In spite of the setbacks, being on dry land marked a new phase of our investigation, one where we would drive across the valley to interview the people most affected by the coming Site C dam. Of course our vehicle was still 100 kilometres away in Hudson's Hope, so we convinced a taxi company to send a car to the boat launch; a $60 cab fare later, we checked in to a hotel in the much closer Fort St. John. We would figure out how to get the truck later.

Located north of where the Site C dam will cross the Peace, Fort St. John is a boom-and-bust logistical staging area and base camp for the masses of contractors and supply industry workers needed to extract BC's natural gas (and now to facilitate

This character was present at an auction in September 2015 just outside of Fort St. John.

the building of Site C). To my eyes it's all angry-looking young men in big pickups, apparently flush with cash.

Over the decades, Fort St. John has metastasized from a farming village into a small city of over twenty thousand.[15] Fort St. John is also suffering the latest downward slide of the resource cycle—the roller coaster that all natural resource–dependent communities must perpetually ride. In the months that follow our visit, the worst oil and gas rout in thirty years plunges western Canada into the doldrums. Gas from the Peace region, particularly the dry stuff from the far north, fetches even lower prices, if you factor in the costs of transporting it to market.[16] Much of the drilling activity here, particularly in places like the Horn River Basin, has fallen off.

We phone up Don Hoffmann, an amateur photographer who was searching for

us on his jet boat the day before, who rallies a charitable network of locals to drive us back to our truck. A petite and feisty woman named Deborah Peck eventually shows up to give us a ride; she and her husband Ross are part of a small, closely knit group of farmers, ranchers and other valley residents who are fighting to save their land from the flood. They have worked as guide outfitters for twenty-five years and now breed horses on the 520 acres they own in the valley, including their pasture and grazing lands spread across multiple parcels between Farrell and Lynx Creeks. They stand to lose significant amounts of land to flooding and sloughing, while facing uncertain impacts to their multiple water supplies from a changing water table and the realignment of Highway 29.

We detour briefly to their home west of Fort St. John on the road to Hudson's Hope. Their expansive multi-level log home, built on 110 acres of river and creek front at the mouth of Farrell Creek by their neighbour Ken Boon, stands on a high bluff overlooking the heavily forested south bank of the Peace River. It's about to become lakefront property, presuming erosion doesn't eventually take the house as well.

Fort St. John has grown into a city of more than twenty thousand people. Its fortunes are dependent on the boom-and-bust cycle of the natural gas industry.
Photograph from Flickr/tuchodi

Left: Ross and Deborah Peck's front porch is decorated with antlers of animals they have hunted in the Peace Valley.

Above: Deborah Peck looks south across the Peace River from her property, which is approximately 17 kilometres east of Hudson's Hope.

Ross and Deborah Peck at home, explaining how a flood will change the Peace Valley.

A rack of colossal elk antlers is perched atop a woodpile by the front door, still connected to bits of meaty red skull cap. Ross shot the animal yesterday just across the river at a spot visible from where we stand; he spent the rest of the day with friends and family hauling 700 pounds of meat up the steep embankment that separates their homestead from the river. "You might say he was at the wrong place at the wrong time," Deborah says of the bull.

In the late 1980s elk were transplanted from the Kootenays to the Dunlevy area on the north side of the Peace arm of the Williston Reservoir, and from there, the population exploded. The surge in numbers has been stoked by the long history of controlled burning focused on south-facing valley slopes, where fire is used as a tool to create ungulate-friendly grasslands. The burning continues to this day to create habitat for elk and Stone's sheep—two species targeted by hunters.

Ross Peck is diminutive but wiry, a quiet man who chooses his words carefully. For him the fight against Site C is deeply personal: his grandfather Ross Darnall owned 160 acres of land at the confluence of the Peace and Carbon Rivers that is now at the bottom of the Williston Reservoir. Darnall battled BC Hydro for years to get more money for his flooded land. By the time the issue was resolved, he was well into his eighties. "My grandfather wasn't against progress, but the trees on that property alone were worth more than BC Hydro offered for the entire thing," said Peck. "The fight broke him in the end."

The prices offered to buy out businesses and landowners by BC Hydro, then and now, are confidential, but resentment over past and future expropriation continues to fester here. Over the last six decades, the threat of Site C has stunted the growth of agriculture and other industries that could have otherwise prospered. Since 1957, the province has held a flood reserve over Crown land lying within the proposed area of the reservoir, which has deterred construction of permanent structures and exploration for oil and gas. Ross pulls out a map of the valley with three giant concentric lines drawn around the river. The smallest loop is a "flood impact line" showing the area that will be destroyed by the Site C reservoir and wind-driven waves. A larger line encircles the estimated extent of inland erosion from the flood, which will be most severe during the first five years of dam operation. The outermost line shows the potential extent of future landslides. "By the time erosion happens here, a cliff will be right on our front deck," says Ross, pointing at his cosy wooden deck chairs. He laughs. "Then I'll just jump."

In a strange twist, this same "flood reserve" boundary has also kept the valley pristine. For the last sixty years, very little oil and gas exploration or mining has occurred within this riparian area. So a line on the map that marks the valley's future destruction and stifles improvement to the land has also halted development that would have otherwise degraded its natural values. Imagine what any of BC's big Okanagan lakes would look like frozen in 1957: this is what makes the Peace Valley so exceptional, in spite of the two existing dams.

The question of how the Site C flood will affect the wildlife here sends Ross into a pensive silence. Moose have already been forced out of the areas north of the river by natural gas extraction, he says, and if Site C destroys what remains of the river and its islands, they will be in trouble. "This country is getting smaller. Oil and gas has wrecked the summer range, the reservoir and new roads will reduce winter range. The impacts are cumulative."

He points to a high distant ridge north of the Peace, part of the Montney natural gas basin, a Greece-sized shale resource that underlies northeast BC and part of Alberta. It's one of two great deposits in the region estimated by the Canadian Association of Petroleum Producers to hold enough natural gas to last three hundred-plus years "at current demand levels"—and much less if BC's liquefied natural gas (LNG) industry for export to Asia ever becomes real. Peck says the ridge being developed north of the river has recently been taken over by Progress Energy; it's part of a wider natural gas play that stretches 200 kilometres north of the river to a place called Pink Mountain.

Progress's parent company is the state-owned Malaysian oil company Petronas, the company that the BC government considers the front-runner to build the first LNG plant on BC tidewater at Prince Rupert. It's one of about twenty plants proposed to process BC natural gas for export to Asia.[17] But despite BC Liberal promises for a great export boom, the economics for the private sector investing in LNG projects have become abysmal: commodity prices have collapsed from a glut of supply, as BC's many global competitors in supplying Asia (notably Australia) have LNG projects that are far more advanced. Such plants are also extremely expensive, and require colossal inputs of energy—one proposed BC plant alone would require 700 MW to cool natural gas to minus 160°C, converting natural gas into a liquid.

Back when LNG seemed like a sure thing, energy analyst and researcher David Hughes found that burning liquefied, fracked gas exported from BC to Asia would ultimately generate emissions similar to coal. He concluded that "exporting BC LNG to China would increase greenhouse gas emissions over at least the next 50 years, compared to building state-of-the-art coal plants." What makes this conclusion astounding is that the LNG export industry has been sold to British Columbia as a great mission to green the climate change–stoking coal-fired plants of Asia.

Virtually all of BC's future natural gas production will involve hydraulic fracking, an energy-intensive process where enormous quantities of water, sand and chemicals are forced under high pressure into horizontal wells to release natural gas trapped in rock. Locals already fear that fracking is affecting a tributary of Lynx Creek north of the Peace, fouling farmers' water supplies. In 2014, the CBC reported that a landslide near a Lynx Creek tributary stream exposed oozing mud and piled silt and soil into the creek; government water testing later showed high levels of lithium, barium and cadmium.

Companies also continue to suck water out of the Peace River and watershed for natural gas extraction. Between January and September 2015 three companies holding water licences removed about 725,000 cubic metres of water from the Peace River watershed—enough to fill nearly 300 Olympic swimming pools—including multiple removals from Williston Reservoir and about 410,000 cubic metres from the lower Peace River itself. These extractions pale in comparison to Texas-based Apache's May 2010 fracking experiment north of Fort Nelson—reportedly the biggest and longest continuous hydraulic fracking operation ever attempted. Water researcher Will Koop reported that over 111 days, from January to April 2010, Apache employed about 900,000 cubic metres of fresh water (about 360 Olympic-sized swimming pools), 111 million pounds of sand and an unknown quantity of toxic "stimulation fluids" to horizontally drill and repeatedly frack at least sixteen wells in the Horn River Basin.

Peck has a simple explanation for why Site C is moving forward now: "On paper there is no strong reason to build the dam. But Site C power will help make BC's liquefied natural gas industry look green."

This statement proves prescient. Up to now it has been presumed that LNG and natural gas extraction would continue to be powered by fossil fuels; in 2012, the province went as far as to change the Clean Energy Act to lift any restrictions on future LNG plants using natural gas to power their operations. But in January of 2016, BC Hydro announced the construction of a $300-million, publicly funded transmission line deep into the Montney gas fields (one of three being planned), to power natural gas operations with BC Hydro grid power. "Before, industrial customers had to burn gas to power their facilities," stated Christy Clark in the press release. "The new transmission line not only makes more projects possible, it means they'll be even cleaner."

Providing BC Hydro grid power to the gas fields is particularly important because the Montney Basin is home to a lot of "wet" natural gas, meaning it contains valuable things like propane, ethane and butane. Especially during the current commodity slump, such spinoff products are more valuable, so gas companies are loath to burn wet gas just to generate energy.

"Irrespective of what happens with LNG, it is in the financial interest of the gas industry to see those operations electrified," says Ben Parfitt, a resource policy analyst with the Canadian Centre for Policy Alternatives. It's this policy of providing clean hydro to the gas industry that provides what Parfitt calls "the only credible explanation for why the Crown corporation is rushing to build the controversial dam at this time."

Deborah has four boxes of freshly harvested potatoes from her garden to deliver before she can drop us off in Hudson's Hope. We head eastward on Highway 29 for our first delivery stop—to the original homestead of Leo Rutledge, the founder of the Peace Valley Environment Association, the community group that mobilized to fight numerous incarnations of Site C. He's been dead for more than a decade, but Rutledge remains a hero in these parts—and a powerful symbol of Site C resistance.

My introduction to Leo Rutledge comes from an unexpected source. Vancouver author Stan Persky came to the Peace in 1985 to write about Site C the last time it was being proposed. (The dam had been defeated in 1983, but showed signs of life into the early 1990s.) He came to this very parcel of riverfront land to research a *This* magazine story called "Keeping the Peace." It was here that he met Rutledge, seventy-four at that point, who had first come to Hudson's Hope by riverboat in 1929. He never left, becoming in turn a trapper, big game hunting guide and grain farmer. By the time Persky showed up, Rutledge's grandson was farming his land. I later talked with Persky in Vancouver, now seventy-five and a university philosophy professor, to ask about the visit. "Leo took me out to the Bennett Dam, showed me the Williston impoundment, and said, 'That's where I used to trap.' He was pointing to the middle of the lake."

Persky vividly remembered Rutledge touring him through Fort St. John, a boom town that was at that point suffering its latest bust. The jobs created to build the W.A.C. Bennett and Peace River Dams had come and gone. Persky's vision of Fort St. John in 1985 remains a premonition of the inevitable future once the Site C stimulus fades and the one-time endowment of natural gas has been liquidated. "Huge sections of the city had vacant houses and motels, nobody was living in them, weeds had grown up in the yards, windows were boarded over," he recalled. "Leo called it 'a ghost town before its time.'"

Rutledge was an unlikely environmentalist, but his life working on the land gave him a unique perspective, which—as Persky reported in the feature article—he used to great effect to pick apart the rationale for building Site C: "'Well, it's just a few acres [being lost to the dam],' they say. But that's the way everything goes to hell. It goes to hell incrementally. Somewhere along the way, you've got to draw the line, otherwise you're going to end up with nothing."

Leo Rutledge, founder of the Peace Valley Environment Association, with a trophy Stone's sheep in the late 1950s.
Hudson's Hope Museum and Archives, HHMA2015.Rut.026

Rutledge was contemptuous of two things: the chamber of commerce boosters who would get behind any development on a matter of principle, and the locals who sold their land to BC Hydro without a fight. Many farmers and landowners in the valley had by that point already agreed to sell. "There was a lot of resentment on the part of people like Rutledge toward the people who had sold out in advance," Persky recalled. "Whereas he hadn't."

More than three decades on, Rutledge is long gone, Site C is back, and I am standing on the same homestead farm fronting the Peace River. Something has profoundly changed, though: about a month ago, Rutledge's surviving children sold the land to BC Hydro. Within days, a bulldozer destroyed the house, which was still fit for habitation. Deborah Peck says this is standard BC Hydro practice, part of an unwritten "scorched-earth policy" to clear the way for the dam, an approach employed for more than half a century in this valley.

The one concession made by Hydro is that a log-cabin-style barn, built by Rutledge in the 1940s with old-growth spruce harvested and hauled up from a river island, is being removed and reassembled at the Hudson's Hope Museum. It's at the

Right: Justin McKnight helps to dismantle Leo Rutledge's log-cabin-styled barn in preparation for it to be moved to the Hudson's Hope Museum grounds. McKnight's nearby home will be flooded by the Site C reservoir.

work party for this relocation that Deborah drops potatoes for several locals who are in the process of dismantling the barn. One of them is Ken Boon, a third-generation farmer, trapper and log cabin builder (he built the Pecks' home) and Leo Rutledge's heir apparent as the ringleader of the Site C landowner resistance.

Boon's head pops up from behind a massive squared spruce log. Compact and wiry strong, the man has a warmth that makes him instantly likeable. He has met Ben before and jokes that he had to leave work the day before to search for our "dead bodies on the river." Ben introduces me as a freelance writer. "Do you think you can sell our story to the *Onion*?"

Deborah is distributing potatoes to various members of the work party, including a long-time local named Dave Kyllo. He lost his river taxi business when the W.A.C. Bennett Dam destroyed the lower Finlay River and has lived with the threat of Site C his entire adult life. Now in his seventies, Kyllo is still vital. He shoves me out of the way when I try to carry potatoes to his car.

Opposite bottom: Ken Boon at work dismantling Leo Rutledge's barn. A farmer, trapper and log cabin builder, Boon is also Rutledge's heir apparent as the ringleader of the Site C landowner resistance.

The loss of Rutledge's waterfront land to BC Hydro is a big psychological blow to the people assembled here. "One generation fights tooth and nail, the next one throws up their hands, and you can't blame them," says Deborah Peck. "We have such a small voice here, and we always will."

The final stop on the potato run is Caroline Beam's place, located on the bank of the river. This home is the most exposed to the river of any that we see. Everything they have will be submerged. Beam's great-grandmother Elizabeth was a White Russian, an ethnic German whose family fled persecution in Tsarist Russia. She married an England-born Peace pioneer named Jim Beattie, who immigrated to the Peace in 1912 after his rise as a concert pianist was derailed by a freak accident (the story goes, he was bucked off a horse and sliced the tendons across his wrist when he landed on a scythe). They raised eight children at their ranch at Gold Bar, about 40 kilometres above the Peace Canyon. It was a wonder for its time—the first homestead to have hot running water, with acres of orchards, vegetables and many head of cattle. The Beattie family would eventually be forced off their property to make way for the W.A.C. Bennett Dam. "My great-grandparents were just told, 'We're taking your land, this is when you have to be out, this is what we will give you for it,'" said Beam. Half a century later, Caroline and her family are facing the same prospect. "My great-grandparents' place was destroyed, then my grandma moved up the river from us,

Above: Elizabeth Beattie in 1967, the last time she ever saw Gold Bar Ranch. This photo was taken the day before the ranch was burned to the ground in advance of the flood that created Williston Reservoir.
Hudson's Hope Museum and Archives, HHMA1981. Bea.003

Above right: The Beatties' Gold Bar Ranch in 1967, about 40 kilometres upstream of the present-day W.A.C.Bennett Dam.
Hudson's Hope Museum and Archives, HHMA1981. Bea.001

Right: Jim Beattie takes a break from working his trapline, 1943.
Hudson's Hope Museum and Archives, HHMA1981. Bea.007

and she fought the last Site C venture. So it's an ongoing thing for us."

Back on the road, we pass the sign for Hudson's Hope, population 1,200. The slogan printed on the sign is "The Playground of the Peace," which is new, having replaced the older tagline: "Of Dinosaurs and Dams." The change reflects two facts: the municipality of Hudson's Hope is divided over Site C and all of the best Cretaceous Period dinosaur remains in the region were drowned when the Peace Canyon Dam (the second dam on the Peace) was completed in 1980. In a backhanded acknowledgment of the loss, the reservoir was named "Dinosaur Lake."

Hudson's Hope is a nondescript frontier town with no main street, established as a fur trading outpost in the early nineteenth century. It has remained on the fringes ever since: railways came too late, and river steamboats were a lifeline here right into the twentieth century. Hudson's Hope was connected by road when the Halfway River Bridge was built in 1938, and the route remained a narrow gravel road for many years; it was partially upgraded to haul coal out of the area in the late 1940s.

The Hudson's Hope we encounter is a tense, conflicted place: it is the town that will be most affected by the flood, but the majority of its residents rely on BC Hydro for a living. After Deborah drops us at our truck, we decide to grab lunch. The diner is packed with members of the transient Gore-Tex consulting intelligentsia. We eat and then drive back to Fort St. John, where we discover our canoe as we had left it, lashed to an alder tree with a bike lock.

Caroline Beam and her children Xavier, Lucas and Tristan at their home on the banks of the river with the Gates pictured in the background. The Beam children have grown up with the river as their backyard.

We wanted to learn more about farmland in the Peace region, and many locals advised us to visit Art Hadland, a local Fort St. John farmer and former Peace River Regional District director. I ring to let him know we're on our way. "Are you the boys who got lost on the river?" he bellows into the receiver, erupting in laughter.

On our way to Hadland's farm north of Fort St. John, we drive through lush farmland growing mostly grains, as well as forage crops for livestock. In addition to cattle and bison ranching, today the Peace district grows more oats, canola, wheat, barley,

Right: We abandoned our rental canoe in the bush at Taylor's Peace Island Park, locking it to a tree with a bike lock. The next day, we were relieved to find it exactly as we had left it.

Below: Cattle belonging to the Ardill Ranch near the Peace River in February 2015.

seed crops and hay than any other part of BC. We pass a field where a single man works a team of horses dragging an old-school plough, its shiny blades half buried in the black earth.

Much of the prosperous farmland we see this day is protected—on paper at least—by BC's Agricultural Land Reserve (ALR), which was created in 1973 as a means to preserve the food-growing capacity of farmland. The reserve was initially a response to growing losses of farmland around the fertile lower Fraser Valley near Vancouver, which by the early seventies was losing more than 4,000 hectares a year of prime farmland to mostly real estate development. The ALR remains today the single most

Cattle belonging to the Ardill Ranch are moved down the road during the seasonal cattle drive. Members of the community come together to help the Ardills with the event.

Long-time rancher Renee Ardill feeds her cattle on her family's ranch in the Peace River valley. Renee's grandparents moved to the area in 1920 and started ranching on the banks of the Peace River along Highway 29 between the Halfway River and Farrell Creek.

notable achievement of Dave Barrett's early 1970s NDP government (which defeated the Socreds and ended W.A.C. Bennett's political career), and is an innovation studied and emulated around the world. It is defended by food security advocates and foodies, and loathed by developers and land speculators province wide.

Upon my return to Vancouver, I had a long conversation with Harold Steves, a lifelong farmer and politician, who is widely regarded as the driving force behind the reserve's creation. He is a steadfast opponent of Site C because it will consume thousands of acres of farmland he considers too valuable to lose. Since the ALR was created in 1973, he tells me, the percentage of fruit and vegetables grown and consumed in British Columbia has dropped from 86 percent to 43 percent today—exactly by half.

The story of the ALR begins around 1959, Steves said, when a new bridge connecting Vancouver to the agricultural community of Richmond prompted the rezoning of thousands of acres of farmland, including the Steves family's dairy farm. One by one, the farmers sold their farms. "The big problem was that nobody knew you could fight city hall in those days, or knew how." Steves's father, after another property tax increase and a crushing court defeat, would eventually sell most of his land too. (Harold still farms what's left of the original 43 acres.)

Harold Steves, the father of the ALR, at home on his farm in Richmond, BC, feeding a Belted Galloway bull named Allen.

Photograph courtesy of Kathy Steves

The loss of the family farm, his birthright, profoundly affected Steves, then a young man studying agriculture at the University of British Columbia. He joined the fledgling New Democratic Party and organized a farming committee in the early 1960s. At one of the meetings, Earl Mattenly, then a village blacksmith in Whalley (in the present-day Vancouver suburb of Surrey), suggested establishing a land bank, something akin to what existed in Saskatchewan. Steves investigated and brought forward a resolution to do something similar in BC. He was on the NDP executive that drafted the 1972 election policy, making a land reserve the number one priority for agriculture. "Then we got elected, so we did it."

Instead of going around and trying to buy up all the farmland in BC, the NDP instead made farmland protection a zoning exercise. Land already zoned for agricultural use went into the new ALR, as in Richmond, where the remaining farmland went into the reserve. Where agricultural zoning did not yet exist, regional districts were tasked with creating agricultural zoning based on such sources as the Canada Land Inventory, which had been conducted since the 1960s to identify lands of agricultural value. The newly protected lands would include large tracts of rich alluvial soils along the Peace River.

Eventually the Peace River Regional District would hold nearly 30 percent of the entire land reserve (about 1.3 million hectares of a total 4.6 million hectares). Much of this agricultural land is clustered today around population centres like Dawson Creek (which is home to just under 11,000 people) and Fort St. John.

On Art Hadland's 2,400-acre seed farm, he produces pedigreed seed not for food, but for sale to other growers. Hadland also served as a commissioner on the North Panel of the Agricultural Land Commission from 1990 to 1996. He tells me Site C will require the single largest removal of farmland from the ALR in the province's history, which would usually require a drawn-out, nightmarish quantity of red tape. Instead, the BC government has simply excluded the land by cabinet order. (Steves confirms the legislation is subject to cabinet decree, meaning the government of the day has the power to set the legislation aside.)

Hadland and his wife Laurel play host for the next four hours. Before we know it, he's barbequing pork chops. At one point he emerges with a briefcase full of Stone Age tools he has found on his land over the years—including hide scrapers and cutting tools that are still sharp ten thousand–plus years later. A natural history buff, he shows us a map of Glacial Lake Peace—a giant water body created by dams of ice and covering the entire site of the Peace Valley after the last ice age—which corresponds with the current boundaries of the ALR. The organic nutrients deposited by Glacial Lake Peace were so rich, the region's best soil falls within the outline of the long-lost super lake.

Opposite bottom left: Art Hadland's briefcase of Clovis tools, including hide scrapers and cutting tools that are still sharp.

Bottom right: Art Hadland, a farmer and former local politician, at home just north of Fort St. John. He says the Site C dam will require the largest exclusion of farmland from the Agricultural Land Reserve in history.

Art Hadland voices a concern that I hear repeatedly during my travels: Site C will be the final nail in the coffin for BC's publicly owned BC Hydro. It is being run into the ground by design. The idea sounds like a conspiracy theory to me, but I also recognize that the government has earned this level of distrust and cynicism. That's because beginning in 2002, the year the new Liberal government launched its first energy plan, Premier Gordon Campbell set about privatizing pieces of BC Hydro. First about 1,500 employees were outsourced to Accenture in a $1.45-billion, ten-year deal that was supposed to save the utility $250 million. Then the transmission arm of the company was cut up and removed, becoming the BC Transmission Corporation (the company was eventually brought back into BC Hydro). Both schemes lost millions.

Consider, says Hadland, that BC Hydro's current debt is over $70 billion. And while Site C is projected to cost about $8.8 billion (including contingency), the recently completed Shepherd Energy facility in Calgary—which produces energy and

Left: Art Hadland shows off an ancient tool he found on his land near Fort St. John. It is at least ten thousand years old.

capacity comparable to that projected for Site C—cost about $1.4 billion to build. By the time the table is cleared, Hadland is tying it all together: the added debt of Site C will break publicly owned BC Hydro, and companies like General Electric and SNC-Lavalin are ready and waiting in the wings. BC Hydro will be broken up into pieces, with all the profitable components—worth many billions and financed for generations by British Columbians—scooped up at fire-sale prices.

It's a sobering end to the night. The next time I see Hadland is less than four months later, not in person, but on the home page of Fort St. John online news site Energeticcity.ca, being handcuffed by an RCMP officer at a Site C protest camp on the edge of the Peace River.

Art Hadland is arrested for mischief on January 6, 2016, near a Site C construction site on the Peace River.
Photograph by Montana Cumming, courtesy of EnergeticCity.ca

I later bounce these ideas off Simon Fraser University energy economist Mark Jaccard, who in the early nineties served as chair of the BC Utilities Commission. One constant throughout his career has been the open question of Site C, kept alive by multiple resurrections. He wrote a master's thesis on it, moved to France, and by the time he returned, it had lurched back to life. Jaccard insists that Site C could be a valuable asset for BC Hydro—eventually. Somewhere down the road, the province will need and use the energy and capacity. The project's value lies in an ability to hold water behind the dam, which can be released at will. He isn't so worried about the debt either, and he doesn't think BC Hydro is in financial trouble. "You never look at debt," he said. "You look at cost of capital, or debt relative to the present value of the revenue stream. And on that basis, BC Hydro is incredible."

He adds, a bit facetiously I think, that the immense value of BC Hydro would be confirmed if we ever went ahead and broke it up, selling it off in pieces to the private sector. "You would find that what people are willing to pay for BC Hydro would be enormous relative to the debt."

Economist Marvin Shaffer agrees that BC Hydro remains a "strong company," but he notes that its rates remain relatively low, which presents a big problem. The BC government proudly points out at any opportunity that British Columbia residents pay the third-lowest residential electricity rates in North America, but this by necessity must change. In 2013 BC Hydro unveiled its forecast of electricity price increases for the coming decade: rates are rising 20 percent in real terms (adjusted for inflation) over the first five years and as much as 30 to 32 percent in real terms over the next ten years.

Rates are going to skyrocket because we need to refurbish our aging hydro dams and other assets—the source of our historically low rates—and develop new energy and capacity to fuel growth. We are also going to have to pay for a lot of bad decisions, says Shaffer. We've committed to buy private power that is costing BC Hydro much more than it is worth; smart meters will end up costing at least $830 million; and the Northwest Transmission Line (NTL) Project, from initial estimates of under $400 million, including the Iskut extension, came in at about $860 million. Moreover, the NTL—which was, like Site C, exempted from BCUC review—will eventually attract mining customers who will pay a fraction of what it costs BC Hydro to develop the new supply and power their industrial operations. "Somebody's got to pay for all of that."

All of these costs will not, however, cause the collapse of BC Hydro, says Shaffer. But they do mark the end of W.A.C. Bennett's era of cheap energy, once known as "the BC Advantage." As Shaffer sees it, "What [the debt] is setting up is a trajectory for fairly heavy rate increases into the future. That would be a real problem, if you're a utility with very high rates, and you try to increase them, and start losing sales."

Land clearing along the Stewart-Cassiar Highway south of Dease Lake in 2012, for BC Hydro's Northwest Transmission Line (NTL). Way over budget and off the public radar, the NTL, like Site C, was exempted from BC Utilities Commission review.
Photograph by Christopher Pollon

And beginning in 2024, the year the dam is expected to be complete, we will begin to pay for something else: the real cost of building Site C. That's because all the billions spent to build it are capitalized. That is to say, they are not included in the rate base, but once the dam becomes operational, that's when customers will start to pay for it. Our future selves and, more to the point, our children will be facing additional double-digit rate increases beginning in or shortly after 2024.

DAY 6 | CONFRONTING THE PEACE INDUSTRIAL SACRIFICE ZONE

OUTSIDE THE Sportsman's Inn in Hudson's Hope, a motel that has not changed its decor since the last Peace dam was built in the late 1970s, a scruffy man in baggy track pants asks me about the canoe on our roof. I tell him we've just spent four days paddling the Peace. "Paddle it while there's still a river to paddle," he spits. "They're even killing protesters over it."

He's referring to the mysterious death of forty-eight-year-old James Daniel McIntyre in July 2015; McIntyre was shot by local RCMP outside a Site C open house in nearby Dawson Creek. He was later

Right: A "Paddle for the Peace" hat sits in the front of a jet boat on the Peace River. This annual fundraiser and awareness event, organized by the Peace Valley Environment Association and West Moberly First Nations, draws together opponents of the Site C dam each July.

PADDLE
FOR THE
PEACE

7th Annual
PADDLE FOR THE
PEACE
2012
Forever Our Peace

confirmed to be affiliated with the hacktivist collective known as Anonymous, a loosely organized cyber-protest group known for hacking industry and government websites. McIntyre was wearing the uniform of Anonymous—a white "Guy Fawkes" mask—and allegedly refusing to drop a knife when he was shot dead on the sidewalk outside the event.

"We will most certainly avenge one of our own when they are cut down in the streets while protesting the earth-wrecking environmental policies of the Canadian government," read a July 18, 2015, statement credited to the group. "Behind this mask there is more than flesh and blood. Behind this mask is an idea and ideas are bulletproof."

Little is known about Anonymous, whose organizational structure has been compared to that of silver-haired hamadryas baboons. (The animals live as part of a larger collective, but form independent splinter groups based around a single dominant male, often acting independently of the wider group.) The victim was known to be opposed to the dam, but the connection between the open house and the shooting was not clear. The RCMP had been called initially because a protester inside the event was causing a disturbance; when the police arrived, they encountered a different man—the masked McIntyre—outside. Almost a year later the Independent Investigations Office of BC, a civilian-led body that investigates police incidents that result in death or serious injury, remains silent.

As the investigation dragged on, I filed a Freedom of Information (FOI) request to BC Hydro, probing if the utility had any involvement with the Anonymous police investigation. "We are unable to confirm or deny the existence of the records you have requested" was the response after a long delay. (A subsequent request seeking details of the internal discussions that led to the above statement came back

Opposite: Kids play by the river's edge at a fundraiser held for the Peace Valley Environment Association, on the edge of Ken and Arlene Boon's alfalfa field.

heavily redacted.) Separate FOI records released by the City of Dawson Creek indicated that Anonymous had gotten revenge—if you can call it that. On August 11, 2015, a denial of service attack crippled and crashed the city website, weeks after the main RCMP website was attacked and disabled for an afternoon. An Anonymous operative claiming to know McIntyre took credit for the Dawson Creek hack and, in a profanity-laden post on social media, threatened to "take from that cop [who shot McIntyre] what he loves the most."

There are people in the Peace, like the man we met in the parking lot, who to this day believe the RCMP deliberately shot and killed James McIntyre for his opposition to Site C. The episode put an immediate chill on peaceful protest against the project, according to Arlene Boon. After the shooting, security was ramped up, and many people in the Peace felt wary of attending Site C protest events. Boon says BC Hydro used the shooting as an excuse to limit the number of open houses and afterward, whenever BC Hydro was out in the valley doing work after the shooting (like on road realignment, for example), they had a security detail in tow.

I am talking with Boon on the edge of her recently cut alfalfa field—a stretch of rich, black soil on the banks of the Peace River near her home. Ben and I are gathered here with about forty other people as part of an annual fundraiser for the Peace Valley Environment Association, the group once led by Leo Rutledge. It's an upbeat crowd. We mingle with a few random journos in attendance, including reporters from the *Alaska Highway News* and CJOB TV in Dawson Creek.

This agricultural land is like a little piece of California set in the BC north, minus the state's devastating multi-year drought. In a graded rating system from Class 1 to Class 7 (where the classes indicate the degree of limitation imposed by the soil in its use for agriculture), it is rated Class 2. There are very few limits to what can be grown here. Farmland like this is more than just good soil: the Peace River valley is a unique east–west conduit for warm ocean air, in a region otherwise dominated by cold Arctic air. That partly explains why the bottom lands of the valley can grow fruit and vegetables usually produced farther south. (One visitor to the region in 1873, a botanist, compared the growing season and conditions here to that of Belleville, a southern Ontario farming centre near Toronto.)

It's here that I meet Mike Van Zandwyk, an aspiring chef and farmer who has just rented an eight-acre plot of land adjacent to this field. Mike's in his early twenties, and the cartilage folds of his left ear are pierced multiple times by the same hooked earring (he looks like the victim of a terrible sport-fishing mishap). He has come to the valley from the Okanagan with his partner Bess to farm before it's too late. "This is the breadbasket of northern BC," he says with a sweep of the hand. "There's not a lot of people growing food here right now. We'd like to change that." Investing time and energy in farming this land is an act of protest for aspiring farmers like Van Zandwyk, in recognition that Site C will impact close to 10,000 acres (an area roughly equivalent to ten Stanley Parks) of Class 1 to Class 5 agricultural land. "With Site C, the jobs and money will come and go. Then what? If more people farmed here, we could build jobs without Site C."[18]

If you want to anger any of the locals assembled on this field, bring up BC Hydro's dismissal of Peace Valley farmland as underutilized. Not only has BC Hydro been buying out farmers here for decades, but investment in and improvements to Class 1 and 2 agricultural lands have been virtually impossible given the proposed dam. That's because the flood reserve was imposed in 1957 by an order-in-council on riverside Crown land, and has remained in force ever since. And for privately owned land, banks are not keen to loan money to improve land that may or may not be under water in a few years. That explains what locals here call the "shadow of the dam": vast tracts of good farmland growing mostly low-value crops like hay for animal feed, or nothing at all. As the locals explain, when farmland comes up for sale here, it is sold at a discounted rate. As a consequence, BC Hydro has been able to snap up such land parcels at a fraction of their true value. In some cases the utility then turns around and leases the land back to a handful of farmers. The vast majority of people still bothering to farm here are people like the Boons, who inherited their land—and with it the threat of Site C.

Underutilized as the land may currently be, there is no denying its productivity potential. Local farmers Blane and MaryAnn Meek, for example, reported growing 82 bushels per acre of canola in 2011, on flats owned by BC Hydro adjacent to Wilder Creek. The recent Alberta provincial average, the highest otherwise in western Canada, was 36.4 bushels. Yet even the Joint Review Panel that scrutinized Site C ultimately agreed with BC Hydro that the permanent loss of the valley-bottom farmland here would not be significant in the wider context of agricultural production in BC or western Canada. In summer 2016 I asked former Joint Review Panel chair

Harry Swain why they came to this conclusion. "Largely because the panel believed in international trade," he said. "We're not going to have self-sufficiency in food in BC." He added that farmers downstream of Site C, outside of the shadow of the dam, were also not growing a lot of vegetables and fruit. "Why isn't that land used for growing vegetables to feed a million people? Well, because you can't make a living doing that."

It's a fact that in the Peace region today, most people are dependent on fruit and vegetables trucked in from Alberta and California. Province-wide, about 60 and 70 percent of BC's imported fruits and vegetables respectively come from the US, with most grown in California, which has long been feeding us. But the value of all the high-quality, underutilized agricultural land we have in the province will only grow in value as the capacity of the southwestern United States and Mexico diminishes, particularly from the parching effects of drought. "Agriculture in California is vulnerable to predicted impacts of climate change, including less reliable water supplies, increased temperatures, and increased pests," reads the state government's California Climate Change agriculture home page. The state, which is reeling under the worst drought on record, has seen farmers (and many other water users) drilling ever deeper for water, to the point that in 2015 at least 60 percent of California's water consumption came from such deep-drilled wells. A 2015 report by the Pacific Institute probing the impacts of the latest drought found that current levels of groundwater use in California far exceeds the natural rate of recharge. The result, say the authors, has been a "decline of groundwater levels across large parts of the state, saltwater intrusion and other water-quality impairments, land subsidence, lost storage, and increased energy costs, among other adverse impacts."[19]

Taking the long view, it seems like common sense to ensure that none of our finite, prime farmland is destroyed unnecessarily, regardless of how much food is being grown there at this moment in time.

To this day, the dismissal of Peace Valley farmland as insignificant makes agrologist Wendy Holm angry. She says BC Hydro excluded over 5,300 acres out of nearly 10,000 acres of unimproved Class 1 to 5 lands in its cost-benefit analysis for the dam. The utility maintained that certain farmland will never be farmed due to inaccessibility and other factors, and that the farmland destroyed makes up a tiny percentage of the total farmland in the Peace. For her part, Holm has calculated that the farmland destroyed by Site C is capable of producing sufficient fresh vegetables to meet the nutritional needs of a million people.[20]

This is the most widely repeated statistic about Peace farmland loss among Site C opponents, and it has been a lightning rod for dissent. But the statistic has also been widely misquoted: the thirty-six-year-old report it is drawn from looked at "fresh vegetable requirements," but dam opponents routinely state that Site C will destroy land that could have produced enough food, period, to feed a million people.

The autumn wind is frigid on this exposed riverfront farm, and people are starting to leave. I notice for the first time a man with a notepad interviewing a woman from the local Saulteau First Nation. He's incongruous among this wider group of farmers and ranchers, wearing a short Indiana Jones leather jacket and a pressed collared shirt, his side-parted hair fluttering in the wind. He's sort of a swashbuckler, impervious to the cold and Gore-Tex fashion trends in these parts. He's definitely not from around here.

I walk over and introduce myself. It's unexpected to meet Amnesty International's Craig Benjamin, human rights campaigner for indigenous peoples, in this part of the world, but the more we talk, the more it computes. He's interested in Site C but is also trying to gauge the wider impact of the boom and bust economy on local people, especially First Nations, as this region is developed for gas, forestry and hydro power. Do we really need to develop thousands of direct jobs from Site C, he asks me, in a region that during the panel review process had an unemployment rate of 3.6 percent? And if the answer is yes, then who exactly are we creating the jobs for?

These are not new questions in a province dependent on resource extraction across its hinterland. Where the big surprise lies is that no one in the Peace country—not the RCMP, not even municipal governments—knows for sure how many workers have descended upon the region at any given time. BC Hydro estimates that Site C alone will create 10,000 direct construction jobs, with about 33,000 jobs through all development and construction stages; a separate Hydro web page claims 7,650 direct construction jobs and up to 35,000 positions through all project stages. Either way, it all sounds good to the current provincial leadership, which won the last election with a vague plan to export LNG and create jobs for working families. But it's not music to everyone's ears, particularly up here. Benjamin says that during the boom times many local people working in the service sector will not share in the benefits and will be forced to absorb the inflated costs of food and housing. When the boom goes away, it will be even worse.

Benjamin is also interested in people like the Boons, who now appear certain to be forced off their land by Site C. Canada is a signatory to a number of legally binding international agreements designed to set stringent rules dictating when citizens can be removed from their lands. He is convinced that BC is in violation of these rules in its approach to Site C. For example, to evict valley residents from their land, international law requires due process, a guarantee of adequate compensation and the recognition that the dam is necessary.

Ken and Arlene Boon at home near the banks of the Peace River.

When I met Arlene Boon in her field, I asked her about the looming threat of expropriation. She seemed strangely upbeat given what they had to lose—and the fact that Site C had already by autumn 2015 become like a steamroller with a lot of forward momentum. "BC Hydro told my grandpa, Lloyd Bentley Senior, that he had to leave from this very spot," she said with a smile. "We haven't even thought about losing or having to move."

But eight months later, a lot has changed. BC Hydro has informed the Boons that they plan to realign the highway right through their homestead, which stands just north of where we stood for the fundraiser. And they want to be building the road

Ken Boon walks with his dog outside of his home and farm on the Peace River near Fort St. John.

by the spring of 2017, which means BC Hydro wants the Boons out before Christmas 2016. "The road goes almost right through our house," Ken Boon told me in June 2016. "We haven't gotten into talking money yet."

Earlier in the year, Ken Boon was in Ottawa, where he met with the UN's special rapporteur on adequate housing to discuss the expropriation of lands due to Site C. The rapporteur's role includes intervening directly and speaking to national governments, based on specific allegations of human rights violations regarding the right to housing (as embodied in the United Nation's International Covenant on Economic, Social and Cultural Rights and other standards and agreements). Boon says he's hopeful that the special rapporteur can help them, but as of this writing, a formal complaint has yet to be lodged by the landowners.

On paper the standards around evicting landowners are very clear, says Benjamin: BC Hydro cannot displace people like the Boons lightly—it has to be absolutely necessary—and alternatives must be exhausted before you go forward. "With Site C, strict necessity is not apparent," he said. "There are alternatives that have not been explored here."

Opposite: Ken Boon in one of his old log homes on his family's farm.

To date the provincial government's response to criticism of Site C, constructive and otherwise, has been to push back. Setting the stage for the pending 2017 election, Christy Clark has launched an attack on "the forces of no," painting opponents of LNG, Site C and other development as enemies of progress, jobs and the revenue needed to fund public schools and hospitals. "There are people who just say no to everything, and heaven knows there are plenty of those in British Columbia," said Clark in early 2016 in response to a First Nations–led coalition formed to oppose an LNG facility proposed near Prince Rupert. "I'm not sure what science the forces of no bring together up there except that it's not really about the science," she said. "It's not really about the fish. It's just about trying to say no. It's about fear of change. It's about a fear of the future."

Against such divisive rhetoric, the opposition New Democrats, the only sizeable political voice of opposition in the BC Legislature, have been unable to articulate a coherent public position on the dam. On one side the party faces its union constituency on a project that promises tens of thousands of construction and spinoff jobs; on the other, they threaten to alienate supporters opposed to Site C's unprecedented environmental impact and the current government's disregard for First Nations rights.

Above: Ninety-one-year-old Vernon Ruskin, the former BC Electric director of planning, who oversaw the original design and planning of the hydro dams on the Peace, including Site C.

Right: Heavy rains in July 2016 caused slope failures on Site C construction access roads. Locals warn that these kinds of slide events on unstable soils will happen again as Site C construction progresses.

Photograph courtesy of Don Hoffmann

With an election less than a year away and construction under way, the NDP has been hesitant to stand fully in favour or against the dam—which has been a gift to the ruling Liberals. Up to now, racing forward with construction to reach "the point of no return" as Christy Clark has termed it—the point when the dam is so far advanced that nothing can stop it—has been nothing short of a cakewalk.

On the question of why the government is so determined to move the project forward under the current timelines, there may be a simple explanation. Could Site C be the most expensive make-work project western Canada has ever seen? Something akin to what Franklin D. Roosevelt did during the Great Depression, when he oversaw the mobilization of sixteen thousand workers to build the Hoover Dam on the Colorado River? Maybe the prospect of creating thirty thousand short-term jobs is enough for the Liberals to justify moving the project forward.

This thought is little consolation to generations of ratepayers facing a nearly $9-billion price tag—which will almost certainly end up much higher. The current Site C cost estimate is particularly confounding to Vernon Ruskin, an engineer and BC Electric director of planning who in the early 1960s oversaw the design and planning for the three Peace River dams: Site A (the W.A.C. Bennett Dam), Site B (the Peace Canyon Dam) and Site C.[21] He's now ninety-one years old, and I met him in May 2016 at his regular perch at the Royal Vancouver Yacht Club in Vancouver. His memory of the work that went in to designing the original Site C was vivid. The original Site C design was a conventional, parabolic-shaped dam, which BC Hydro has since redesigned into a right-angle dam. The result is an unnecessary surge in cost for no apparent benefit, he said, other than to put more money in consultants' pockets. "Now it takes more concrete, more rock fill, and so on," he said. "It's also weaker. So we have to reinforce it more for earthquakes."

But what bugged Ruskin the most involved Site C's cost estimate: they announced the total project cost (including a $440-million contingency) in advance of having contractors bid on it. "When [contractors] see an estimate of seven billion dollars, they're not going to give you a price for less than seven billion dollars!"

It's our last night in the Peace Country, and we're spending it in Fort St. John. What to do on a Saturday night in a fading boomtown?

We decide to get a first-hand look at the gas fields that sprawl for hundreds of kilometres north of town—the alleged wellspring of BC's future prosperity. We randomly choose a point on the map—a settlement called "Montney" about 25 kilometres north of Fort St. John—named for the vast geological shale gas formation that underlies much of this area. We arrive to little more than an intersection with a boarded up storefront and a few dilapidated houses. We continue north.

It's a shock to drive into British Columbia's oil and gas country and find nothing but bucolic farmer's fields as far as the eye can see. But as we continue north, rural roads give way to fresh logging roads, and before we know it, we are in the midst of the Squirrel gas field, on a site run by a Calgary company called Pengrowth. The well we encounter is an unimpressive structure: a short metal staircase leading up to a drilling platform in the centre of a six-hectare clear-cut. It's unclear whether it is active or abandoned, and the distinction is important. There were at least ten thousand inactive well sites in northeastern British Columbia in 2014, and the situation is much worse in Alberta. In a report submitted to the BC Oil and Gas Commission by the David Suzuki Foundation in late 2015, the current taxpayer clean-up liability from this northeast BC legacy alone could be as much as $700 million.

North of the village of Montney, we explored this natural gas site in the midst of a six-hectare clear-cut.

Orphan wells are just part of the wider cost the entire Peace region must pay to be the economic engine of the province. Call it the resource curse: between 1974 and 2010, almost 70 percent of the Peace River region's five biggest watersheds, encompassing an area bigger than Croatia, had been altered by land use and ecological fragmentation. Consider that since the first commercial production of natural gas in 1948, 1,100 kilometres of transmission lines have been built, and 45,000 kilometres of roads now criss-cross the region, the latter equal in total distance to seven one-way roadtrips between Vancouver and Halifax. If we consider the sum of the impacts from all the gas wells, pipelines, clear-cuts, coal mines and hydro development, the Peace is not so much an economic engine as an expendable, industrial sacrifice zone.

In March of 2015, the Blueberry River First Nations, whose reserve is just north of Fort St. John, decided enough was enough. They filed a lawsuit in BC Supreme Court,

the first lawsuit to zero in on BC's breach of Treaty 8 on the basis of the "sum of destructive industrial impacts of regional development." (The lawsuit is ongoing.) About four years earlier, the Doig River First Nation, whose traditional territory has been degraded mostly by natural gas extraction, took action of a different kind. The nation unilaterally declared the boundaries of K'ih tsaa?dze—a 90,000 hectare "tribal park" straddling the BC and Alberta border about 75 kilometres northeast of Fort St. John.

Such unilateral action has precedents. Logging prompted the Haida Nation to create the Duu Guusd Tribal Park in 1981 on Haida Gwaii, which went on to receive BC government recognition as a protected area In 1984 the Tla-o-qui-aht First Nation established a tribal park near Tofino to protect Meares Island from logging, which remains in force today. More recently, the Tsilhqot'in, coming off a legal victory that confirmed the First Nation's title to a large chunk of its traditional territory, established the 3,000-square-kilometre Dasiqox Tribal Park in fall of 2014; the park included Fish Lake, which up until recently was designated to become a tailings dump for an open pit mine.

Wetlands of K'ih tsaa?dze—part of a 90,000 hectare tribal park straddling the BC and Alberta border about 75 kilometres northeast of Fort St. John.

Photograph courtesy of Herb Hammond

What is particularly important about Doig River's desperate action in the Peace is that the nation is a signatory to Treaty 8, and what they have created at K'ih tsaa?dze attempts to implement a way of managing land that closely mirrors the intent of their ancestors at the moment they first signed the treaty.

It's dark as we drive in silence back to Fort St. John and our hotel, which is connected to a casino. Given the province's high-stakes gambling around Site C and LNG, the latter feels like a fitting venue for our last few hours in town. I lose at slots as Ben hits the card tables, drinking doubles and getting giddy as his winnings mount.

This place is surprisingly dead for a Saturday night. Ben plays on, while a bartender entertains me with tales of Fort St. John woe. Any given month, he says, half a dozen high-end pickup trucks are abandoned in the hotel's Walmart-sized parking lot by young men who never return. The keys often hang in the ignition. He speculates that drugs, debt or both are to blame. "They just up and leave," he says.

We get up early the next morning and decide it's time to do the same.

POSTSCRIPT

IN THE MONTHS THAT FOLLOW our return from the Peace, the Liberal government of Christy Clark has taken to scheming about where surplus Site C energy might go. One option is to supply Alberta with clean hydro by tapping federal infrastructure funds to improve east-west transmission access—to offset the province's coal-fired thermal generation and, more specifically, to green the Alberta oil sands. "Site C could be a key part of establishing Canada's brand around the world for both oil and gas as the cleanest produced anywhere on the globe," said Clark in April 2016, confirming that discussions had already started between the two provinces. But establishing new transmission infrastructure will be expensive and complicated, and, without a large price imposed on carbon, it's unlikely that hydro would be more economical to Alberta oil sands operations than gas-fired co-generation. For its part, Alberta has made it clear that it will not consider buying surplus BC power unless it is able to move forward with any number of contentious plans to pipe oil sands bitumen across BC to the Pacific coast, for tanker export to Asia.

Aerial shot of K'ih tsaa?dze—a 90,000 hectare tribal park straddling the BC and Alberta border.

Photograph courtesy of Herb Hammond

Clark's quest to find a use for surplus Site C electricity—months after construction has started—would be comical if I wasn't one of the ratepayers on the hook to eventually pay for it. It's especially preposterous-sounding to Marc Eliesen, former BC Hydro CEO, who sounds as though he is hyperventilating over the phone from Whistler when I bring up the subject.

Believe it or not, he tells me, there is a Canadian precedent for the situation with Site C. Beginning in the early 1960s, Manitoba set about planning to build a series of hydro generating stations on the Nelson River, projects even larger than Site C, including one called Limestone. It broke ground in 1976, based on Manitoba Hydro forecasts that showed the power would be needed ten years later. "Two years into the construction, they recognized that all their long-term forecasts were going out the window," says Eliesen. "And they stopped." While the project was on hold, the utility went to work and eventually negotiated a 500 MW $2.2 billion power sale to Minnesota. Eliesen knows all of this because he was the chairman of the Manitoba Hydro-Electric Board from 1984 to 1988, and one of his principal tasks was to determine whether or not there would be customers for the electricity. "So with that contract [in hand] we said, 'Okay, we can now start building Limestone.' We went to the National Energy Board, we held hearings, we had regulatory approval. And the Limestone Generating Station was built."

With Site C, it's just the opposite, says Eliesen. Heavy industrial power users are slowing down, forcing the revision of BC Hydro's longer-term forecasts. But instead of stopping and negotiating with potential export customers from a position of strength, the BC Liberal job-creation juggernaut just keeps rolling forward. "Site C is slated, in my judgment, to become a white elephant," says Eliesen. "Which means you've got roughly ten billion dollars [of] additional burden on the ratepayers of BC Hydro."

Not that it needs to be that way. In travelling up to the Peace, our initial plan was to ignore the politics and be agnostic. Keep it simple. But anthropologists we are not, and for me, Limestone is the kicker. This project was already well on the way to becoming a white elephant when it was stopped; Manitoba Hydro had already gone as far as building a coffer dam—a temporary enclosure that would allow the actual dam to be built on dry riverbed, more than two full years into the construction schedule—when the decision was made. Construction resumed in 1985, almost seven years after it had been suspended, with the first unit coming online in 1990 and the station being completed in 1992. All it took to make all of this happen was leadership.

Like the energy Site C may one day generate, the dam is invisible to most British Columbians, located in a far-off industrialized hinterland most will never see or even consider. So for many, the cost only becomes real when we think about what we won't be able to buy as a consequence of building this single project. Economists call these foregone alternatives the "opportunity cost." Consider that for roughly the same amount of money British Columbia could pay for seismic upgrading for high-risk schools, a Broadway corridor subway line and the new St. Paul's Hospital in Vancouver, and the planned replacement of the George Massey Tunnel in the Fraser Valley with a ten-lane super bridge, with enough left over to build a treatment plant for Victoria's sewage and a few brand new, intermediate-class ferries thrown in. And that's just what might happen in the southwestern corner of the province, where about half of the population is concentrated.

Instead, Site C has become for the provincial government an opportunity, something that the LNG export industry could never be. Instead of waiting for a risk-averse private sector and Asian markets to materialize, here is a mega project with the premier's finger securely on the trigger. The Agricultural Land Reserve? Mandatory Utilities Commission scrutiny? Treaty rights and real consultation with First Nations? Not a problem: just channel the bullish spirit of W.A.C. Bennett and build the damn thing!

By late summer 2016, almost a year after we returned from our explorations of the river, the race to advance Site C beyond the point of no return has become ever more urgent for the provincial government. At the same time, it has become abundantly clear that Site C, no matter how badly conceived or executed, will not be the end of the world for us. Ben and I returned to our busy lives in the city, where the lights flick easily on and off, and the earth beneath our feet, barring the next mega-thrust earthquake, is not going anywhere soon. But we cannot unlearn what we know: we've shared tiny island sanctuaries with bull elk, coyotes, beavers and bats, and searched in vain for the ghosts of long-dead fur traders and explorers. We have stood on a riverside alfalfa field with farmers Arlene and Ken Boon, whose nearby generations-old homestead will soon have a highway realigned through the front door. We've witnessed scenes of natural abundance that (we fancy) would have shocked Alexander Mackenzie, and wandered the fringes of a vast shale gas sacrifice zone. We have also looked uncomfortably on as a young Saulteau First Nation woman teared up in frustration as she tried to explain the gravity of seeing her ancestors' unmarked graves disappear to the bottom of a fifty-metre-deep reservoir.

With each passing day now, this stretch of land and water moves closer to becoming a "landscape on the verge of an apocalyptic flood." It's a phrase I found so alluring, before I understood the complicity inherent in writing the last sentence of a book about Site C, on a computer enabled by electricity generated by the Peace River.

SELECT BIBLIOGRAPHY

Abbey, Edward. *Desert Solitaire: A Season in the Wilderness*. New York: Ballantine, 1968.

Anderson, Margaret Sequin. *The Tsimshian: Images of the Past, Views for the Present*. Vancouver: UBC Press, 1984.

Ansar, Atif, et al. "Should We Build More Large Dams? The Actual Costs of Hydropower Megaproject Development." *Energy Policy*, March 2014, pp. 1–14.

Apps, Clayton. *Assessing Cumulative Impacts to Wide-Ranging Species Across the Peace Break Region of Northeastern British Columbia.* Final Report, Version 3.0. Aspen Wildlife Research Inc., June 2013.

Baker, Paula. "Two Hunters Seriously Injured After Grizzly Bear Attack South of Fort Nelson." *Global News*, September 7, 2015. Accessed at globalnews.ca.

Bein, Sierra. "Everything We Know About the Death of an Anonymous 'Comrade' in RCMP Shooting." *Vice*, July 20, 2015. Accessed at www.vice.com.

Bowes, Gordon E., ed. *Peace River Chronicles.* Vancouver: Prescott Publishing, 1963.

BC Hydro. "Environmental Impact Statement Executive Summary," amended July 19, 2013. Accessed at bchydro.com.

BC Hydro. "Environmental Impact Statement Fact Sheet, Environmental Background," Volume 2, Section 11. Accessed at bchydro.com.

BC Hydro. "New transmission line to power development in south Peace," press release. January 20, 2016. Accessed at bchydro.com.

BC Liberal Party. "Strong Economy Secure Tomorrow," 2013 Election Platform. Accessed at www.poltext.org.

BC Oil and Gas Commission. "Quarterly Oil and Gas Water Use Summary, July–September 2015." Accessed at www.bcogc.ca.

BC Utilities Commission. *Site C Report in the Matter of the Application of British Columbia Hydro and Power Authority for an Energy Project Certificate for the Peace River Site C Project. Report & Recommendations to the Lieutenant Governor-in-Council.* May 1983. Accessed at www.sitecproject.com.

Burley, D.V., et al. *Prophecy of the Swan, The Upper Peace River Fur Trade of 1794–1823*. Vancouver: UBC Press, 1996.

Caldicott, Arthur. "BC Hydro's $76 Billion Debt." *Watershed Sentinel*, March 1, 2016. Accessed at www.watershedsentinel.ca.

California Energy Commission. *Commission Guidebook, Renewables Portfolio Standard Eligibility,* eighth edition, June 2015. Available athttp://www.energy.ca.gov/renewables/tracking_progress/documents/renewable.pdf.

Chan, Cheryl. "'The Bear Hunted Him': B.C. Camper Killed by Black Bear Didn't Stand a Chance," *The National Post*, May 12, 2015. Accessed at news.nationalpost.com.

Opposite: A mule deer on the banks of the Peace River during winter.

Concerned Scholars. "Briefing Note #1: First Nations and Site C." Prepared in support of *Site C: Statement by Concerned Scholars*, May 24, 2016. Accessed at https://sitecstatement.org.

Concerned Scholars. "Briefing Note #2: Assessing Alternatives to Site C (Environmental Effects Comparison)." Prepared in support of *Site C: Statement by Concerned Scholars*, May 24, 2016. Accessed at https://sitecstatement.org.

Concerned Scholars. "Briefing Note #3: The Regulatory Process for the Site C Project." Prepared in support of *Site C: Statement by Concerned Scholars*, May 24, 2016. Accessed at https://sitecstatement.org.

Cooley, H., et al. *Impacts of California's Ongoing Drought: Agriculture.* The Pacific Institute, August 2015. Available at http://pacinst.org/app/uploads/2015/08/ImpactsOnCaliforniaDrought-Ag.pdf.

Cox, Sarah. "BC Hydro Apologizes for Bennett Dam's 'Profound and Painful' Impact on First Nations at Gallery Opening." *DeSmog Canada*, June 10, 2016. Accessed at desmog.ca.

Dickey, James. *Deliverance.* New York: Bantam Doubleday Dell, 1970.

ERM Consultants Canada. *Aboriginal Health Risk Assessment of Mercury in Bull Trout Harvested from the Crooked River, British Columbia.* Prepared for the West Moberly First Nations, Vancouver, British Columbia, March 2015.

Feinstein, Asa. *BC's Peace River Valley and Climate Change: The Role of the Valley's Forests and Agricultural Land in Climate Change Mitigation and Adaptation.* Chillborne Environmental, February 2010.

Fisheries and Oceans Canada. *Technical Review of the Effects of the Site C Clean Energy Project on Fish and Fish Habitat of the Peace River, British Columbia.* Canadian Science Advisory Secretariat, Science Response 2014/011. Accessed at www.dfo-mpo.gc.ca.

Fumoleau, René. *As Long as This Land Shall Last: A History of Treaty 8 and Treaty 11, 1870–1939.* Calgary: University of Calgary Press, 2004.

Garfield, Brad. *Bear vs. Man: Recent Attacks and How to Avoid the Increasing Danger.* Minocqua, WI: Willow Creek Press, 2001.

Gilchrist, Emma. "'Dereliction of Duty': Chair of Site C Panel on B.C.'s Failure to Investigate Alternatives to Mega Dam." *DeSmog Canada*, March 11, 2015. Accessed at www.desmog.ca.

———. "'It's No Longer About Saying No': How B.C.'s First Nations Are Taking Charge With Tribal Parks." *DeSmog Canada*, March 29, 2016. Accessed at www.desmog.ca.

Global Forest Watch Canada and the David Suzuki Foundation. *Passages from the Peace: Community Reflections on BC's Changing Peace Region.* October 2013. Accessed at www.davidsuzuki.org.

Harden, Blaine. *A River Lost: The Life and Death of the Columbia.* 2nd ed., New York: W.W. Norton & Company, 2012.

Hoagland, Edward. *Notes from the Century Before: A Journal from British Columbia.* New York: Random House, 1969. Revised ed., 2002.

Hughes, David. *A Clear Look at BC LNG: Energy Security, Environmental Implications and Economic Potential.* Canadian Centre for Policy Alternatives, May 26, 2015.

Hume, Mark. "Huge 'Tribal Park' With Stunning Ecosystem to Span B.C.–Alberta," *The Globe and Mail*, October 2, 2011.

Hunter, J. "B.C. First Nation Says Legacy of Dam Construction has Poisoned Trout," *The Globe and Mail*, May 11, 2015.

IUCN Red List of Threatened Species. "Ursus americanus." Accessed at iucnredlist.org.

IUCN Red List of Threatened Species. "Ursus arctos." Accessed at iucnredlist.org.

Joint Review Panel, Site C Clean Energy Project. *Report of the Joint Review Panel: Site C Clean Energy Project*. Her Majesty the Queen in Right of Canada, May 1, 2014. Accessed at ceaa-acee.gc.ca.

Joint Review Panel, Site C Clean Energy Project."Report Summary: 50 Recommendations," *Report of the Joint Review Panel: Site C Clean Energy Project*. Her Majesty the Queen in Right of Canada, May 1, 2014. Accessed at ceaa-acee.gc.ca.

Keystone Wildlife Research, *Peace River Site C Hydro Project Stage 2 Baseline Vegetation and Wildlife Report.* Report prepared for BC Hydro, 2008–2009. Accessed at www.env.gov.bc.ca.

Koop, Will. "The World's Biggest Experimental Frack Job! (Fracking 24/7... Fracking All Night Long.)" BC Tap Water Alliance, June 17, 2010. Accessed at www.bctwa.org.

Lee, Peter G., and Matt Hanneman. "Atlas of Land Cover, Industrial Land Uses and Industrial-caused Land Changes in the Peace Region of British Columbia." Global Forest Watch, 2013. Accessed at www.davidsuzuki.org.

Lewis, A. David and mpMann. *Some New Kind of Slaughter: Or Lost in the Flood (and How We Found Home Again)*. Chicago: Archaia, 2007.

Matheson, Shirlee Smith, and Earl K. Pollon. *This Was Our Valley.* 3rd ed. Detselig Enterprises, 2003.

Meissner, D. "B.C. Premier Christy Clark Strikes Back at LNG Opponents." *CBC News*, January 26, 2016. Accessed at www.cbc.ca.

Nathwani, J. "Canada's Low Carbon Electricity Advantage: Unlocking the Potential of Inter-Regional Trade," in *Canada: Becoming a Sustainable Energy Superpower*, ed. R. Marceau, Canadian Academy of Engineering, May/June 2014, pp. 101–102.

Neering, Rosemary. *W.A.C. Bennett.* Don Mills, ON: Fitzhenry & Whiteside, 1981.

Parfitt, Ben. "Ever Wondered Why Site C Rhymes With LNG?" *DeSmog Canada*, February 4, 2016. Accessed at www.desmog.ca.

Persky, Stan. "Keeping the Peace: A Site C Memoir." Dooneyscafe.com, March 2, 2016. (Originally published in *This* magazine, 1985.) Accessed at dooneyscafe.com.

Province of British Columbia, Technical Briefing Presentation by Energy and Mines Minister Bill Bennett, December 17, 2014. Accessed at www.youtube.com.

Rana Law. "K'ih tsaa?dze Tribal Park." Accessed at http://www.ranalaw.com/kih-tsaadze-tribal-park.

Royal Society of Canada, and 250 Canadian Scholars. "250 Canadian University Professors and Royal Society of Canada Speaking Out Against Site C Mega Dam." News release, May 24, 2016. Accessed at newswire.ca.

Seymour, Garrett. Transcript of a spoken-word video played at Aatse Davie School, Kwadacha First Nation, Fort Ware, British Columbia, on December 5, 2006, for the Kemess North Copper-Gold Mine Project. Joint Review Panel Hearings conducted pursuant to the Canadian Environmental Assessment Act, Proceedings at Hearing Volume XII.

Shaffer, Marvin. *Assessment of the Need for and Alternatives to the Site C Project.* Report prepared for the Peace Valley Environment Association for submission to the CEAA and BCEAO Joint Review Panel. November 25, 2013.

Sherman, Paddy. *Bennett.* Toronto: McClelland and Stewart, 1966.

Sherwood, Jay. *Surveying Northern British Columbia: A Photojournal of Frank Swannell.* Prince George: Caitlin Press, 2004, pp. 127–150.

Smith, James K. *Alexander Mackenzie.* Don Mills: Fitzhenry & Whiteside, 1976.

The Society for Pollution and Environmental Control. *The Case Against Site C.* Vancouver, 1983.

Stantec Consulting. *Site C Clean Energy Project: Greenhouse Gases Technical Report.* Report prepared for BC Hydro, Project No. 121810787, December 6, 2012.

Stodalka, William. "Site C Dam Will Not Be Diverted to BC Utilities Commission." *Alaska Highway News,* October 5, 2015. Accessed at www.alaskahighwaynews.ca.

Swain, Harry. "Opinion: Site C: Truly Awful Economics." *The Vancouver Sun*, June 16, 2016. Accessed at www.vancouversun.ca.

Swannell, F.C. "Ninety Years Later." *The Beaver*, Spring 1956, pp. 32–37.

Tomblin, Jordan, and Greg Jenion. "Sentencing 'Anonymous': Exacerbating the Civil Divide Between Online Citizens and Government." *Police Practice and Research*, July 8, 2016.

Trumpener, Betsy. "Fracking, Landslide Blamed for Contamination of Northern BC Creek." *CBC News*, October 25, 2015. Available at www.cbc.ca.

Wakefield, Jonny. "Justice Minister, Courts Won't Derail Site C, Christy Clark Says." *Alaska Highway News*, April 7, 2016. Accessed at alaskahighwaynews.ca

Wilson, Michael C. "Late Quaternary Vertebrates and the Opening of the Ice-Free Corridor, with Special Reference to the Genus Bison," *Quaternary International*, 32 (1996).

The World Commission on Dams. *Dams and Development: A New Framework for Decision-Making*, November 2000. Accessed at www.unep.org.

Zamora, Amanda, et al. "California's Drought Is Part of a Much Bigger Water Crisis. Here's What You Need to Know." From the series *Killing the Colorado*. *ProPublica*, June 25, 2015. Accessed at www.propublica.org.

Museum curator Elinor Morrissey prepares to clean artifacts that are showcased in the Hudson's Hope Museum.

ENDNOTES

1. All brown bears and grizzlies are the same species, although some biologists disagree on whether there are distinct subspecies or ecotypes; according to bear expert Jeff Gailus, the common term "grizzly bear" refers to *Ursus arctos* that occur in interior and noncoastal environments, while "brown bear" is used to refer to coastal individuals or populations, and those occurring in Eurasia.

2. In 2008 and 2009, BC Hydro announced that the Kwadacha and Tsay Keh Dene First Nations reached respective settlements with BC Hydro and the province over litigation brought by the groups stemming from the construction of the W.A.C. Bennett Dam.

3. According to historian Gordon E. Bowes, if it had not been for the discovery of gold in 1861 and 1862 along the banks of the Parsnip and Peace Rivers, the Rocky Mountains—not the 120th meridian—would have formed the northeast provincial boundary between Alberta and BC. "British Columbia would have lost its great wheat and petroleum areas to Alberta," Bowes writes.

Opposite: Caroline Beam east of Hudson's Hope on the Peace River.

4. The reasons laid out were myriad: the destruction of habitat for bull trout and other food fish; the mercury levels in fish already experienced as a result of the past Peace dams and the anticipated mercury spike to come with Site C; the destruction of "ancient, sacred sites and graves of our ancestors"; and the loss of farmland and calving areas for moose, mule deer and elk.

5. Such a mitigation could be also used by Arctic grayling, mountain whitefish, rainbow trout and other big fish as they nose up against the Site C dam. It is not clear, however, how successful such mitigations will ultimately be.

6. His explorations extended the North West Company's fur trade business into the western fringes of the Peace River region with the establishment of Rocky Mountain Fort, the first non–First Nations settlement in the interior of what would become British Columbia, not far from present-day Fort St. John.

7. An eight-kilometre section of highway that hugs the river along Bear Flats will have to be rerouted northward away from the flood waters and erosion of the new reservoir, taking at least five homes with it.

8. According to Michael Church of UBC, the entire river, excluding the alpine headwaters, is within Canada's boreal zone. But it's confusing: in terms of national, generalized "natural regions," Church told me, the Peace River lies entirely within the boreal forest. In more refined regional terms, in BC the Peace River lies in the "sub-boreal spruce zone."

9. After 1823, the fur trade ceased altogether for almost forty years in these parts, after Beaver First Nation warriors massacred eight Hudson's Bay Company staff at St. John, a nearby fur trade post. At least one contemporary source suggested the killings were provoked by fur trader Samuel Black's fondness for stealing wives from the local Native men. A more recent explanation is that the Beaver were enraged at plans to move St. John far away, forcing them to travel regularly through the territory of hostile rivals.

10. On the causes of these cost overruns, study co-author Flyvbjerg is absolutely scathing. In a press release that accompanied the February 2013 study, he grouped forecast-making experts into two camps: "fools" and "liars." "Fools are reckless optimists who see the future with rose-tinted glasses," he stated. "These forecasting fools ignore hard facts and uncertainty, betting the family silver on gambles with very low probability of success. Liars deliberately mislead the public for private gain, fiscal or political, by painting overly-positive prospects of an investment, just to get it going."

11. Modelling for total emissions, including from construction and construction materials, over the 108-year life of the project, envisioned a number of different scenarios, including "likely" and "conservative" emissions of 5.3 and 7.3 million tonnes respectively of CO_2 equivalent, according to a December 2012 BC Hydro consultant's report.

12. The panel also had a mandate to "examine proposals for the mitigation of adverse effects, and to record assertions of Project effects on the Aboriginal rights and treaty rights of the affected First Nations and Métis peoples."

13. Sold to the public as "green power," IPPs such as Plutonic Power and AltaGas set about building huge energy projects, including East Toba/Montrose and Forrest Kerr, which both approached 200-MW capacity and required major river-altering construction works and a spider web of new roads and transmission line rights-of-way.

14. Under Section 5 of the Utilities Commission Act the Utilities Commission could perform a review of the project, providing unbinding "advice" to government. While that was happening, the RSC called on the federal government to revisit the order-in-council approving the project, and to probe whether or not Site C infringes upon aboriginal and treaty rights. Until court cases are settled, both governments were called upon to hold back on issuing permits and authorizations needed for construction to proceed.

15. As of July 6, 2016, the single family homes sold so far this year in Fort St. John had a median selling price of $387,000.

16. The Henry Hub spot price for natural gas, the US price of gas, was over $13 per million BTU in June 2008; when we were in the Peace region in September, it had collapsed to about $2.60. As I type these words in September 2016, it is up to about $3.00.

17. "It's no fantasy," reads the BC Liberal 2013 election platform in reference to the nascent LNG export industry in BC. "We can create $1 trillion in economic activity and create the BC Prosperity Fund with $100 billion over 30 years." The platform predicted LNG would create thirty-nine thousand construction jobs in BC, with another 75,000 full-time jobs once in operation.

18. By summer 2016, Mike and Bess had begun selling their vegetables at farmers' markets and operating a highway market stall at Bear Flats.

19. Beyond the California drought, the state of the wider Colorado Basin—which supplies water to about 40 million people across California and six other states, supporting 15 percent of the US food supply—is a grave concern. ProPublica has reported that human mismanagement of this Basin, including large water withdrawals for agriculture, "has left the West more vulnerable to both short and long-term changes in climate."

20. Holm based her "million people" calculation on a 1980 consultant report from BC Hydro, which found that "1,100 acres of alluvial soils could meet the fresh vegetable requirements of roughly 266,000."

21. Sites D and E were originally envisioned as well. In May 2016, BC Hydro confirmed that "Site D was eliminated as an option by BC Hydro in 1967. There will be no Site E."

A truck descends Highway 29 at Bear Flats. The rising flood will require a realignment of sections of road in this area, destroying five homes.

INDEX

Bold indicates a photo or map

Opposite: A collage of Polaroid images Ben took during our canoe trip in the Peace River valley.

ABOUT THE AUTHORS

Christopher Pollon (twitter: @C_Pollon) is a Vancouver-based freelance journalist who reports on business and the politics of natural resources, focusing on energy, mines and oceans. His work has appeared in *National Geographic Books*, *Reader's Digest*, *The Globe and Mail*, and *The Walrus*. He is a contributing editor at *The Tyee*.

Ben Nelms is a Vancouver freelance photojournalist whose work has appeared in *The New York Times*, *Maclean's*, *The Globe and Mail*, *Sports Illustrated* and *Canadian Geographic*. His portfolio can be viewed online at www.bennelms.ca.